Thomas Minchin Goodeve

The elements of mechanism

Fourth Edition

Thomas Minchin Goodeve

The elements of mechanism
Fourth Edition

ISBN/EAN: 9783337278168

Printed in Europe, USA, Canada, Australia, Japan

Cover: Foto ©berggeist007 / pixelio.de

More available books at **www.hansebooks.com**

THE ELEMENTS

OF

MECHANISM.

DESIGNED for STUDENTS of APPLIED MECHANICS.

BY

T. M. GOODEVE, M.A.

Barrister-at-Law:
Lecturer on Applied Mechanics at the Royal School of Mines.

FOURTH EDITION.

LONDON:
LONGMANS, GREEN, AND CO.
1874.

All rights reserved.

PREFACE.

TWO EDITIONS of this Treatise have already been published as a separate and independent work. It has been rewritten and enlarged, in order to provide an elementary text-book on the principles of mechanism, suited to the series of Elementary Text-Books on Mechanical and Physical Science now in course of publication by Messrs. Longman & Co.

The intention of the Author has been to afford some aid in the systematic preparation which a student must pass through in order to understand the intricate combinations of modern machinery, and it is hoped that a statement of some of the leading principles which govern the applications of mechanism may be found useful by many who are engaged in occupations connected with applied mechanics.

TEMPLE : *October* 1870.

CONTENTS.

INTRODUCTION.

General Statement. Definitions. Spur and Bevil Wheels. Belts and Bands. Guide Pulleys. The Screw Surface. Pitch of a Screw. The Worm Wheel. The Screw Thread . . . PAGE 1

CHAPTER I.

ON THE CONVERSION OF CIRCULAR INTO RECIPROCATING MOTION.

Art. 1—4. Resolution of Motion. 5—10. The Crank and Connecting Rod; the Eccentric Circle. 11. The Swash Plate. 12. Valve Motion. 13—17. Escapements. 18—29. Cams. 30—37. Circular converted into Reciprocating Motion by Wheelwork; various Machines. 38. The use of Crossed and Open Bands. 39—43. Mangle and Segmental Wheels. 44—48. Quick Return Movements. 49—51. The Stanhope Levers; the Knuckle Joint. 52—53. Linkwork 19

CHAPTER II.

ON THE CONVERSION OF RECIPROCATING INTO CIRCULAR MOTION.

Art. 54—55. General Principles. 56—58. Ratchet Wheels. 59—60. Feed Motions. 61. Nipping Lever. 62. Silent Feed. 63. Escapement. 64. Levers of Lagarousse. 65. Screw Barrel . 87

CHAPTER III.

ON THE TEETH OF WHEELS.

Art. 66—72. Statement of the Problem; the Pitch of a Wheel; the Sector; Preliminary Propositions. 73—75. Solution of the Problem.

76—83. Various Forms of Teeth. 84—86. Involute Teeth. 87—90. General Considerations. 91. Bevil Wheels. 92. Transfer of Motion . . . . . . PAGE 101

CHAPTER IV.

ON THE USE OF WHEELS IN TRAINS.

Art. 93—95. Examples of Wheels in Trains; the Copying Principle. 96. The Eight-day Clock. 97—98. The Screw-cutting Lathe. 99. Cranes. 100—103. Connection of Axes . . . 129

CHAPTER V.

ON AGGREGATE MOTION.

Art. 104—108. Examples. 109—111. Epicyclic Trains. 112. Ferguson's Paradox. 113. Sun and Planet Wheels. 114—117. Epicyclic Trains. 118—119. The Cordelier. 120. Equation Clocks. 121—122. Slow Motion. 123—133. Parallel Motion. 134—135. The Indicator. 136—137. Governors of Steam-engines. 138—144. Drilling and Boring Machines. 145. The Oval Chuck 150

CHAPTER VI.

ON MISCELLANEOUS CONTRIVANCES.

Art. 146—154. The Fusee and its Applications; the Snail; the Disc and Roller. 155. The Geneva Stop. 156. The Double Eccentric. 157. Step Wheels. 158. Variable Crank. 159—160. Cone Pulleys. 161. The Star Wheel. 162—164. Rolling Curves. 165—166. Counting and Numbering Machines. 167—174. Parallel Axes; Hooke's Joint. 175. Bell-crank Levers. 176—180. The Escapements of Clocks and Watches . . . 212

ELEMENTS
OF
MECHANISM.

INTRODUCTION.

A MACHINE may be defined to be an assemblage of moving parts, constructed for the purpose of transmitting motion or force, and of modifying, in various ways, the motion or force so transmitted.

The parts of a machine are set in motion by some moving power, which may be derived from any convenient source; and the machine itself must be constructed with reference, not merely to the character of the power from which its motion is derived, but also with a careful regard to the action of the moving parts, and the forces which are capable of executing the work required of it.

' In commencing the systematic study of machinery, it will be readily understood that certain simple relations of motion are traceable between the prime mover which starts the machinery and the pieces which execute the work; and it is also clear that, in practice, relations governing the transmission of force must exist as certainly as those which govern the transmission of motion. The considerations relating to force may often occupy the mind of the mechanic in a greater degree than those which refer to motion; but in reducing the subject to analysis it will be found most convenient to separate entirely the two points of view, and to confine our attention in the first instance mainly to *Theoretical Mechanism*—that is, to an examination of the various contrivances and arrangements of parts in machinery whereby motion is set up or modified, and to disregard or postpone any enquiry into the mechanical laws which control the forces concerned in these movements. But as the present work is intended for general use and study by practical men, the author will to a small extent break through this general rule, and will take occasion, where the enquiry would be useful, of pointing out also the manner in which certain pieces of mechanism have served a compound object in transmitting exact and definite amounts of motion, while dealing at the same time with refined and subtle distinctions as to the method of transmitting force.

The subject of escapements, for example, is one of this complex character: it is impossible to understand why some contrivances have worked more successfully than others in actual practice, without understanding also the nature of the mechanical laws involved in the arrangement of the parts, and it would be a pity to sacrifice, for mere formal reasons, any opportunity which may arise of placing useful information before a student of the subject.

We have now to consider and arrange the method according to which our enquiries are to be carried on, and if we were to pause for a moment and look back upon that rapid creation of machinery which followed so closely upon the splendid invention of the steam-engine by Watt, we should naturally expect that some uniform arrangement for applying steam-power would be adopted by common consent, and that this arrangement would powerfully influence the art of constructive mechanism. Accordingly we find that in applying the power derived from steam for the purpose of driving machinery in our mills and factories, it is the practice to connect the engine with a heavy fly-wheel, the rotation of which is made as uniform as possible, and then to carry on, by lengths of shafting, the uniform motion of this fly-wheel to each individual machine in a factory.

Suppose, for example, that we were visiting a cotton mill, and were examining and endeavouring to comprehend the action of a complete piece of machinery such as a power-loom for weaving calico. We should at once see that every moving piece, acting to produce the required result, derived its motion from the uniform and constant rotation of a disc or pulley, outside the machine itself, and communicating by means of a band with the shafting driven by the engine, and thus it would become obvious that the problem of making a machine resolved itself mainly into a question of the resolution or transfer of circular motion in every variety of manner, and subject to every possible modification.

Accordingly we propose to commence, in the present treatise, with a discussion of the methods adopted for the conversion of circular into reciprocating motion; then to proceed to the reverse problem, or to examine the conversion of reciprocating into circular motion; to pass on to an enquiry into the results deduced from combinations of wheel-work in trains, giving the theory which has led to an accurate shaping of the teeth of wheels; to consider, further, some arrangements by which a moving piece may be made the recipient of two or more independent motions; and finally, to conclude by collecting and analysing certain miscellaneous contrivances which produce results of a specific and very noticeable character.

It will be necessary, however, to premise a few general remarks and definitions.

In the transfer of motion or force from one axis to another, wheels furnished with teeth are commonly employed. The various calculations connected with the forms of teeth which are suitable for this purpose will be given hereafter; at the present time we may remark that the most simple case of the transfer of motion from one axis to another occurs when a circular disc or plate moves another in its own plane by rolling contact.

In such a case the uniform motion of the axis, A, conveys a perfectly even and uniform motion to the other axis, B.

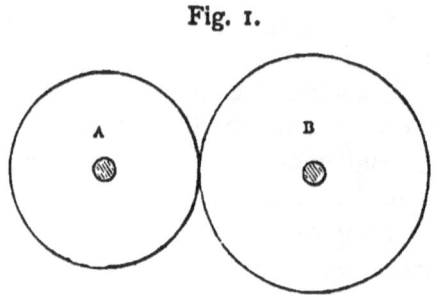

Fig. 1.

If A and B were circular plates with flat edges, and very accurately adjusted, it would be quite possible for A to move B by friction alone, the two plates rolling smoothly and evenly upon each other without any slipping of the surfaces in contact, but we could not expect A to overcome any great resistance to motion in B; or, in other words, we could not convey force by the action of one disc upon the other.

The transmission of force being an essential condition in machinery, the discs A and B are provided with teeth, as in the annexed figure, and the mechanist endeavours so to form and shape the teeth that the motion shall be exactly the same as if one circle rolled upon another.

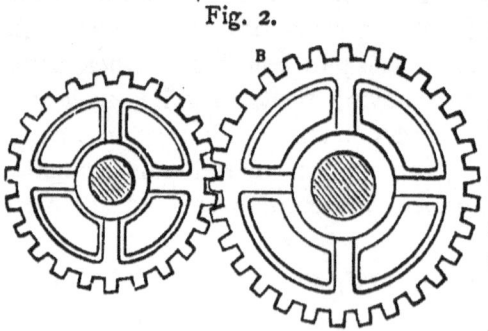

Fig. 2.

Herein consists the perfection of wheelwork, a perfectly uniform motion of the axis A is to be conveyed by means of teeth to the axis B; and the motion of B, when tested with microscopic accuracy, is to be no less even and uniform than that of A.

Since, then, it appears that the motions of A and B are exactly the same as those of two circles rolling upon each other, such imaginary circles may always be conceived to exist, and are called the *pitch circles* of the wheels in question. They are represented by the dotted lines in Fig. 3.

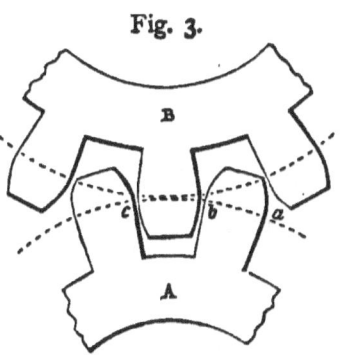

Fig. 3.

The pitch circle of a toothed wheel is an important element, and determines its value in transmitting motion.

Suppose that two axes at a distance of 10 inches are to be connected by wheelwork, and are required to revolve with velocities in the proportion of 3 to 2. Two circles, centred upon the respective axes, and having radii 4 and 6 inches, would, by rolling contact, move with the desired relative velocity, and would, in fact, be the pitch circles of the wheels when made. So that whatever may be the forms of the teeth upon the wheels to be constructed, the pitch circles are determined beforehand, and must have the proportion already stated.

Note.—It is usual to call the fraction $\frac{3}{2}$ the *velocity ratio* between the two axes.

It appears also that when the number of teeth upon a wheel is indefinitely increased, the wheel itself degenerates into the pitch circle.

So much of the tooth as lies within the pitch circle is called its *root* or *flank*, and the portion beyond the pitch circle is called the *point* or *addendum*.

The *pitch of a tooth* is the space $a\ c$ upon the pitch circle cut off by the corresponding edges of two consecutive teeth.

Spur wheels are represented in Fig. 2, and are those in which the teeth project radially along the circumference.

In a *face wheel*, cogs or pins, acting as teeth, are fastened perpendicularly to the plane of the wheel; in a *crown wheel* the teeth are cut upon the edge of a circular band; and *annular wheels* have the teeth formed upon the inside of an annulus or ring, instead of upon the outer circumference.

A straight bar provided with teeth is called a *rack*, and a wheel with a small number of teeth is termed a *pinion*.

The spur wheels, before described, are suited to convey motion only between parallel axes; it often happens, however, that the axes concerned in any movement are not parallel, and as a consequence they may, or may not, meet in a point. If the axes do not intersect we proceed by successive steps, and continually introduce intermediate inter-

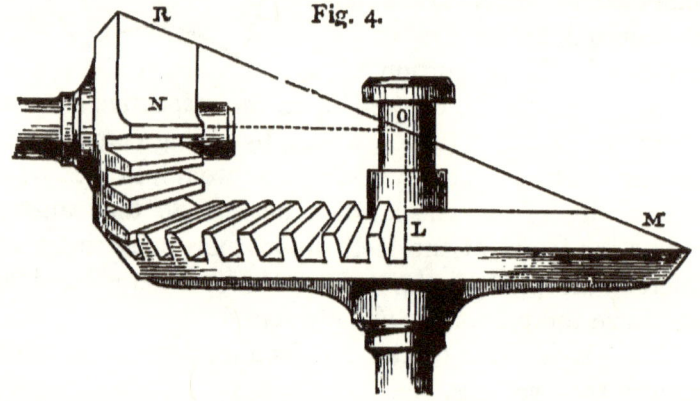

Fig. 4.

secting axes, and thus we are led to the use of inclined wheels whose axes meet each other, and which are known as *Bevil wheels*. (Fig. 4.)

It is easily proved in Geometry that two right cones which have a common vertex will roll upon each other, and the same would be true of the frusta of two cones such as L M and N R, which are represented as having a common vertex in the point O.

The rolling of the cones will allow us to consider any pair of circles in contact and perpendicular to the respective axes as the pitch circles of the frusta, and teeth may accordingly be shaped upon them so as to produce the same even motion as that which exists in the case of spur wheels.

This fact about the rolling of two cones becomes very

clear when enquired into, and it is evident that if one of the cones be flattened out into a plane table, by increasing its vertical angle up to 180°, the property of rolling will not be interfered with. But in that case the common vertex will be a fixed point in the table, and accordingly, if we roll a cone upon a table, the vertex ought not to move in the least degree as the cone runs round.

It is quite easy to test the matter in this way, and if the table be smooth and level the apex will remain perfectly stationary, although the cone itself is free to run in any direction.

The principle under discussion is sometimes applied in the construction of machinery; there is a large circular saw in the Arsenal at Woolwich which is driven by the rolling contact of the frustra of two cones, and upon examination it will be found that the directions of the axes of the two frustra meet exactly in the centre of the revolving circular saw.

Equal bevil wheels whose axes are at right angles are termed *mitre wheels*.

It is sometimes convenient that the axes of the bevil wheels should pass close to each other without intersecting; the teeth have then a twisted form, and the wheels are known as *skew bevils*.

Belts or *straps*, otherwise called *bands*, are much used in machinery, in order to communicate motion between two axes at a distance from each other. In this case an endless band is stretched over the circumference of a disc or pulley upon each axis, and the motion is the same as if the discs rolled directly upon each other. The usual form of the pulley is shown in Fig. 5.

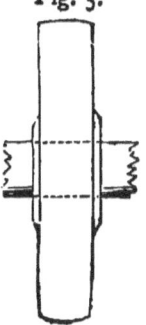

Fig. 5.

It is a common practice to convey steam-power by means of shafting and wheelwork to the various floors of a mill, and then to distribute it to the separate machines by the aid of straps or belts.

These straps adhere by friction to the surfaces of pulleys, and work with a smooth and noiseless action; but they are subject to two principal objections which may or may not be counterbalanced by their other advantages. The friction of the axes upon their bearings is increased by the double pull of the strap arising from its tension, and there is a liability to some change in the exactness of the transmission of motion by the possible stretching or slipping of the band.

It may be useful here to enquire how the necessary size or strength of the strap is ascertained when a given force is transmitted, and we take the following example :—

Suppose that a force of 5 horse-power is to be carried on by a strap moving with a velocity of 600 ft. per minute over a suitable pulley. The work done by 5 horses is $5 \times 33,000$ foot-pounds per minute, and the work done by the strap must be the same.

Let P be the pull upon the strap in pounds, then $P \times 600$ is the work done by the strap in one minute,

$$\therefore P \times 600 = 5 \times 33000$$

$$\therefore P = \frac{5 \times 330}{6} = 5 \times 55 = 275 \text{ lbs.}$$

If the velocity of a point in the strap had been reduced to 300 ft. per minute, P would have been 550 lbs., if it had been increased to 3,000 ft. per minute, P would have been 55 lbs., and thus we recognise the well known mechanical principle, that the slower the movement by which any given force is transmitted, the greater must be the strength with which the moving parts are constructed.

In carrying out this principle, successful attempts have recently been made to convey by means of a slender wire rope, moving at a velocity of some 6000 ft. per minute, the power of even 100 horses to a considerable distance.

The advantage proposed being to utilise the power of water in districts where it is now wasted.

So, again, at the locomotive workshops of the London and

North-Western Railway at Crewe, a cotton rope ⅝ths of an inch in diameter, and weighing about 1½ oz. per foot is carried along the length of a workshop with a velocity of 5000 ft. per minute, and is employed to actuate a traversing crane which is adapted for lifting a weight of 25 tons. The velocity of 5000 ft. per minute is reduced, by suitable mechanism, to that of 1 foot 7½ inches per minute, and the requisite work can be done by subjecting the whole cord to no greater strain than that of 109 lbs.

Here we should point out that the term *band* is applied either to a flat strap or a round cord indifferently. The best material for round bands, such as are used in light machinery, is no doubt catgut, and then the band is fitted with a hook and eye to make it continuous. It must work in a pulley with a grooved rim, or it would slip off, and this groove prevents our shifting it easily from one pulley to another. The power of readily shifting a driving band is often an indispensable condition, and can be obtained at once by the use of a flat band, which will hold on to its pulley with perfect security if we only take care to make the rim slightly convex, as shown in Fig. 5, instead of being concave. No groove is necessary, or indeed admissible; and, upon entering a workshop where steam-power is employed, you see each machine driven by a flat strap riding upon one of a set of two or more pulleys with perfectly smooth edges.

The strap has no tendency to slip off, and it is shifted with the greatest ease from one pulley to another when pressed a little upon the advancing side by a fork suitably placed.

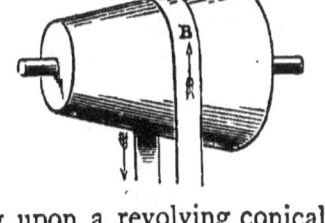
Fig. 6.

The reason of this curious fact will be apparent if we examine the case of a tight belt running upon a revolving conical pulley. The belt embraces the cone, and tends to lie flat upon the slant surface, thus becoming bent into the form

A B, the portion B being somewhat nearer to the base of the cone than the portion A. (Fig. 6.)

The cone, during its revolution, exerts an effort to carry B onward in a circle parallel to its base, and the consequence is that the belt tends to remain upon the slant surface of the cone, and to rise higher rather than to slip off.

In like manner, if a second cone of equal size were fastened to the one shown in the drawing, the bases of the two cones being joined together, the strap would, if its length were properly adjusted, work its way up to the part where the bases met, and would ride securely upon the angular portion formed by their junction; but this is the same case as that of a slightly convex pulley, for it is evident that a little rounding off of the angle at the junction of the bases would convert this portion of the double cone into a convex pulley. Thus the action becomes perfectly intelligible.

The *fast and loose pulley* is an adjunct of the driving belt. It consists of two pulleys, whereof one is keyed to the shaft to which motion is to be conveyed, and the other rides loose upon it; when the strap is shifted from the loose to the fastened pulley the shaft will begin to rotate, otherwise it remains at rest, the loose pulley alone turning round.

The band is shifted by a fork, which is made to press laterally upon its *advancing* side.

The advancing portion of the band must always lie in the plane of the pulley round which it is wrapped, but the retreating portion may be pulled on one side without causing the band to leave the pulley. This rule applies whether the band is round or flat.

It is by observing this condition that a band may be used to communicate motion between two axes which are not parallel and which do not meet in a point.

Problems such as these are interesting, as presenting difficulties to be overcome by a knowledge of principles.

Suppose that we are required to arrange that a band

Straps or Bands.

working over a pulley upon one given axis shall drive another pulley upon an axis at right angles to the first.

Here we intend that the pulleys should be placed one above the other as in the sketch. As the band goes round

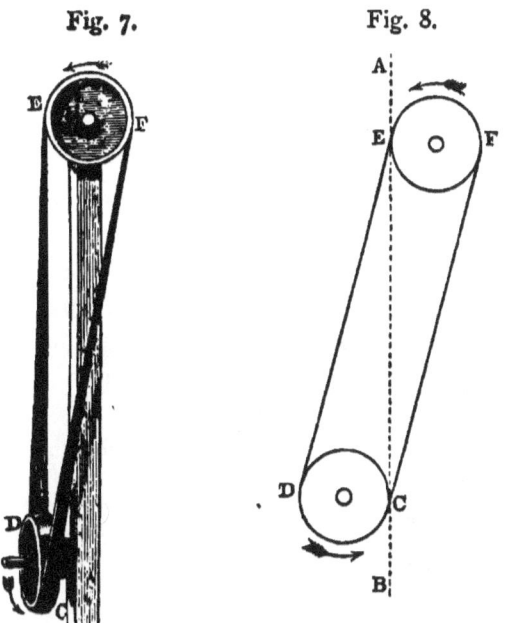

Fig. 7.   Fig. 8.

we have to provide that its advancing portion shall always lie in the plane of the pulley upon which it works. The easiest way of proceeding is to draw a straight line, A B, upon paper and to place circles representing the pulleys in contact with A B upon each side of it.

Draw now the lines E D, C F to represent bands passing round the circles, and however you may bend the two planes containing the pulleys by folding the paper about A B as an edge, it is clear that the advancing portion of the strap will continue to lie in the plane of its pulley so long as the motion occurs in the direction of the arrows.

Reverse the motion and the strap will leave the pulleys at once.

Guide pulleys are sometimes used, and they are constructed as follows :—

Conceive that a band moving in the direction of A B is to be diverted into another direction, C D. There are two cases to consider.

Fig. 9.

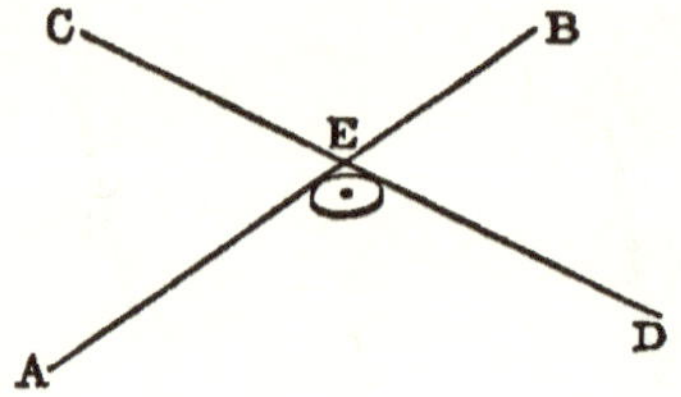

1. Let A B and C D meet in E. At the angle E place a small guide pulley whose plane is coincident with the plane A E D. This pulley obeys the required condition, and will therefore answer its purpose.

2. If A B and C D do not meet, or do not meet within a reasonable distance.

Draw any straight line, E F, cutting both A B and C D. In the plane A E F, and at the angle E, place a guide pulley, E,

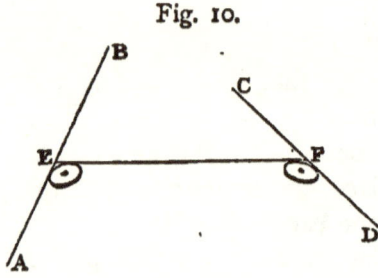

Fig. 10.

and do the same at F by placing a guide pulley in the plane E F D, and thus the strap will be carried on.

One advantage in the use of guide pulleys will be found in the fact that they enable us to overcome the inconvenience of not being able to reverse the motion when the planes of the pulleys are inclined to each other.

Thus, conceive that two pulleys work in the planes z A x,

Z A y inclined to each other at an angle x A y. In the line of the intersection of the planes, viz. A Z, take any two convenient points, H and B, place one guide pulley at H in the plane C H D, and another at B in the plane E B F, then the band C H D F B E will run round the two main pulleys securely in either direction.

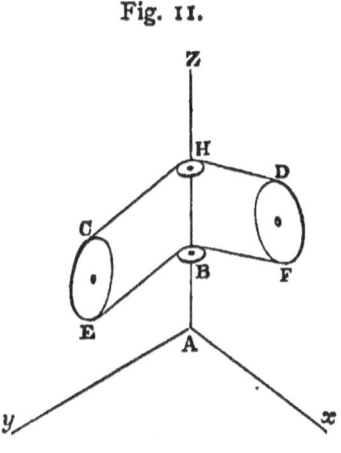

Fig. 11.

This is evident, as we have done nothing to infringe the necessary condition, each advancing and retreating portion of the band will, in both cases, be found in the plane of the pulley upon which it rides.

Instead of bands we may employ chains to communicate motion from one axis to another, and there is one instance where a chain is always so used, viz. in the transfer of the pull of the spring from the barrel to the fusee of a watch. Here the form of chain is the type of most others of the heaviest construction, consisting of one flat plate or link rivetted to two others, which are placed one above and the other below it, and thus the chain consists of one and two plate-links alternately. When a chain of this sort is used to transmit great force, it is called a *gearing chain*, and the open spaces formed by the two parallel links engage with projections on the wheel or disc over which it runs, rendering it impossible for the chain to slip.

One practical objection to the use of chains where great accuracy is required consists in the fact that the links are liable to stretch, and that the pitch or spacing may thereby lose its exactness, the result being to cause some jar and vibration in the working.

After toothed wheels, the screw plays the most important

part in mechanical appliances, and indeed it is difficult to over-estimate its value or utility. The screw bolt and nut are used to unite the various parts of machinery in close and firm contact, and are peculiarly fitted for that purpose; then, again, the screw is employed in the slide rest and in the planing machine to give a smooth and even longitudinal motion, the same purpose for which it aids the astronomer in measuring the last minute intervals which are recognisable in the telescope. In the screw press we rely upon it to transmit force, we use it in screw piles to obtain a firm foundation for piers or lighthouses, and as a propeller for ships it has given a new element of strength and power to our navy.

The definitions relating to the screw are the following.

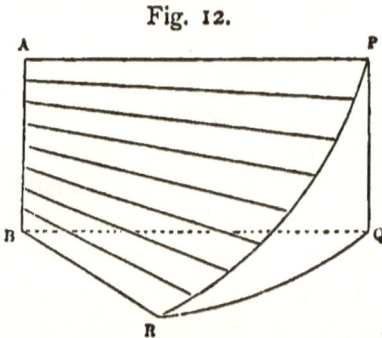

Fig. 12.

If a horizontal line A P, which always passes through a fixed vertical line, be made to revolve uniformly in one direction, and at the same time to ascend or descend with a uniform velocity, it will trace out a *screw surface* A P R B, in the manner indicated in the sketch.

The points of intersection of this generating line with any circular cylinder whose axis coincides with A B, will form a *screw thread*, P R, upon the surface of the cylinder.

The *screw thread* used in machinery is a projecting rim of a certain definite form, running round the cylinder, and obeying the same geometrical law as the ideal thread which we have just described.

The *pitch of a screw* is the space along A B, through which the generating line moves in completing one entire revolution, but in practice the pitch of a screw bolt is usually estimated by observing the number of ridges which occur in an inch of its length; thus we speak of a screw of one-eighth of an inch pitch as being a screw with eight threads to the inch.

Also A B is called the *length* of the screw surface A P R D, and the angle P R Q represents the *angle* of the screw.

In the diagram, A P is shown as describing a *right-handed* screw; if it revolved in the opposite direction during its descent, it would describe a *left-handed* screw.

If a single thread were wound evenly round a cylinder, and the path of a thread marked out, we should have a *single-threaded* screw; whereas, if two parallel threads were wound on side by side, we should obtain a *double-threaded* screw.

The object of increasing the number of threads is to fill up the space which would be unoccupied if a fine thread of rapid pitch were traced upon a bolt, and thus, to give the bolt greater strength in resisting any strain which tends to strip away the thread. Increasing the number of threads makes no difference in the pitch of the screw, which is dependent on any one continuous thread of the combination.

The ordinary screw propeller is a doubled bladed screw, and has sometimes three or even four blades, which correspond to the multiple threads here spoken of.

A *worm wheel* is a wheel furnished with teeth set obliquely upon its rim, and so shaped as to be capable of engaging with the thread of a screw; the revolution of the endless screw or worm A B will then impart rotation to the wheel C, and the wheel will advance through one, two, or three threads, upon each revolution of A B, according as the thread thereon traced is a single, double, or triple thread.

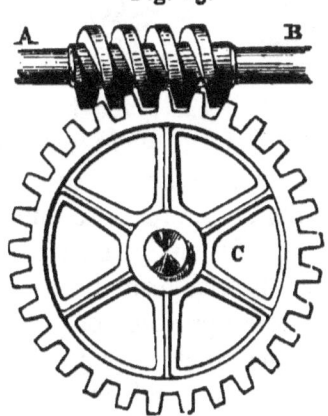

Fig. 13.

This reduction of velocity

causes the combination to be particularly valuable as a simple means of obtaining mechanical advantage, and, as we have stated, the number of threads upon the screw determines the number of teeth by which the wheel will advance during each revolution of A B.

In the transmission of force the screw is always employed to drive the wheel, and necessarily so, because the friction would prevent the possibility of driving the screw by means of the wheel, even if the loss of power were disregarded; but in very light mechanism, where the friction is insensible, the wheel may drive the screw, and then the screw is frequently connected with a revolving fly, and serves to regulate the rate at which a train of wheels terminating in the worm wheel may run round.

The two principal forms of screw-thread used by engineers are the square and the V thread; they are given in the sketch, and in applying them we should understand that there are three essential characters belonging to a screw-thread, viz. its *pitch*, *depth* and *form*; and three principal conditions required in a screw when completed, viz. *power*, *strength*, and *durability*.

Fig. 14.

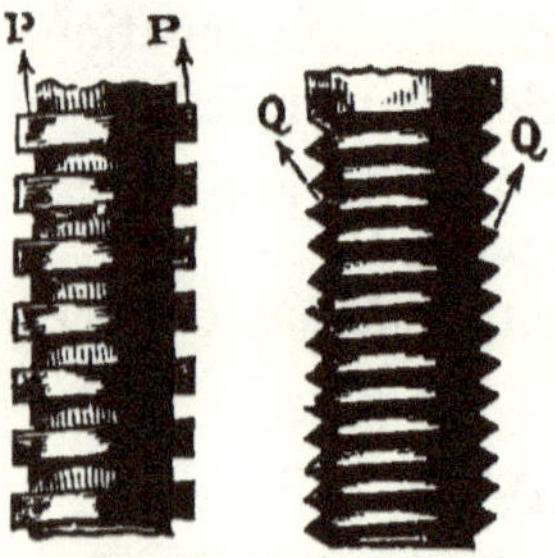

It is easy to see that no one can declare exactly what power, strength, or durability is given by a screw-thread of a certain pitch, depth, or form, when traced out upon a given cylinder. The problem is indeterminate, and must remain so; we cannot lay down any rule for determining the diameter of a screw bolt required for any given purpose, nor can we say what should be the precise form of thread.

It is the province of practical men to determine any such questions when they arise, being guided by experience and by certain general considerations which we propose now to examine.

The Screw-thread.

1. The *power* of a screwed bolt depends upon the *pitch* and *form* of the thread.

If the screw-thread were an ideal line running round a cylinder, the power would depend solely on the pitch, according to the relation given in all books on mechanics, viz.—

weight × pitch = power × $\begin{cases} \text{circumference of the circle de-} \\ \text{scribed by the end of the} \\ \text{lever-handle.} \end{cases}$

If the thread were square we should substitute for the ideal line a small strip of surface, being a portion of the screw surface shown in Fig. 12, which would present a reaction P to the weight or pressure everywhere identical in direction with that which occurs in the case of the ideal thread. Hence, if there were no friction, we should lose nothing by the use of a square thread in the place of a line.

A square-threaded screw is therefore the most powerful of all, and is employed commonly in screw presses.

But if the thread were angular, the reaction Q which supports the weight or pressure would suffer a second deflection from the direction of the axis of the cylinder over and above that due to the pitch, by reason of the dipping of the surface of the angular thread, and we should be throwing away part of the force at our disposal in a useless tendency to burst the nut in which the screw works.

In this sense, the square thread is more powerful than an angular or V thread of the same pitch.

2. The *strength* depends on the *form* and *depth*.

This statement is obvious. In a square thread half the material is cut away, and the resistance to any stripping of the thread must be less than in case of the angular ridges.

Again, if we deepen the thread we lessen the cylinder from which the screw would be torn if it gave way, and thus a deep thread weakens a bolt.

3. Finally, the *durability* of a screw-thread depends chiefly upon its depth, that is, upon the amount of bearing surface; and in the case of a screw which is in constant use, as, for

C

example, in the slide-rest of a lathe, it would be well for the young mechanic to satisfy himself upon this point by ascertaining the amount of bearing surface given by the fine deep thread which is found upon the screw working in the slide-rest of a well-made lathe.

Probably the finest specimen of minute workmanship in screw cutting, will be found in the screws provided by Mr. Simms for moving the cross wires or web across the field of view of a micrometer microscope.

There are 150 threads to the inch, the diameter of the bolt being about $\frac{1}{8}$th of an inch; the head of the screw is a graduated circle read off to 100 parts, and the movement of the wires produced by turning the screw-head through the space of one graduation, is quite apparent.

Upon examining the thread with a microscope, we should see a fine angular screw, consisting of a number of comparatively deep-cut ridges, having the sides a little inclined and the edges rounded off.

We conclude the section by pointing out the meaning of some words which will occur frequently.

The term *axis* denotes the central line of a cylinder, and is a mathematical phrase: an engineer distinguishes a heavy cylindrical piece of metal as *shafting*, or a *shaft*, and designates smaller cylindrical bars as *spindles*; a wheelwright speaks of the *axle* of a wheel, and a watchmaker calls the same thing an *arbor*.

Of two moving pieces, that which transmits motion is termed the *driver*, and that which receives it is the *follower*.

Gearing and *Gear* are the words used to indicate the combination of any number of parts in a machine which are employed for a common object.

Toothed wheels are said to be *in gear* when they are capable of moving each other, and *out of gear* when they are shifted into a position where the teeth cease to act.

CHAPTER I.

ON THE CONVERSION OF CIRCULAR INTO RECIPROCATING MOTION.

Art. 1. The most simple case of motion is that of a point describing a straight line with a uniform velocity. Whenever a point deviates from a rectilinear path, and describes a plane curve of any form, it must be the subject of two independent movements in lines at right angles to each other.

Thus, let a point move in a circular path, A P B, with a uniform velocity, then it is evident that the point, P, while describing the arc A P, has been the subject of two rectilinear movements, one from A to Q, in the direction of the diameter A B, the other from Q to P, in a perpendicular direction.

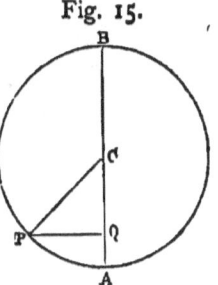

Fig. 15.

In other words, circular motion is of a compound character, and is capable of resolution into its elements; if it be thus resolved, and if one equivalent be suppressed, so that the motion of Q is substituted for that of P, we obtain the fundamental case of the conversion of circular into reciprocating motion.

2. In order to express the relation between the actual positions of the points P and Q in analytical language, we proceed as follows :—

Let C P $= a$, P C A $=$ C.
Then A Q $=$ A C $-$ C Q $=$ A C $-$ C P cos C,
or A Q $= a(1 - \cos \text{C})$.

Ex.—To find the position of C P when Q is half-way between A and C,

Here $AQ = \dfrac{AC}{2} = \dfrac{a}{2}$

$\therefore \dfrac{a}{2} = a(1 - \cos C)$,

whence $1 - \cos C = \dfrac{1}{2}$, $\cos C = 1 - \dfrac{1}{2} = \dfrac{1}{2}$,

or $C = 60°$.

3. In analysing this movement we must first ascertain the relation which exists between the actual velocities of P and Q at any instant.

Suppose C P to sweep round with a uniform velocity, as would commonly be the case, and we see at once that Q begins to move slowly at A, comes to rest gradually at B, and that its greatest speed occurs when just passing through C.

The motion of Q is not uniform, and its rate of advance at any given instant may be found by a method well known to mathematicians, and which we give in a note.*

The result is, that $\dfrac{\text{velocity of Q}}{\text{velocity of P}} = \sin C$.

The sine of C has all values from 0 to 1 which are registered in a table of natural sines, and by substitution from this table we can find how much the velocity of P differs from that of Q at any period of the motion.

When $C = 45°$, $\sin C = \cdot 7071068$,

$\therefore$ vel. of Q $= \dfrac{707}{1000}$ vel. of P, very nearly.

When $C = 90°$, $\sin C = 1$, $\therefore$ vel. of Q = vel. of P, which is evidently true, because the point P is then moving in a direction parallel to A B.

---

*   Let $ACP = \theta$, $CP = a$, t the time from A to P.

Then $AP = a\theta$, $AQ = a(1 - \cos\theta)$

$\therefore$ vel. of P $= \dfrac{d.AP}{dt} = \dfrac{a\,d\theta}{dt}$, vel. of Q $= \dfrac{d.AQ}{dt} = a \sin\theta \dfrac{d\theta}{dt}$

or $\dfrac{\text{vel. of Q}}{\text{vel. of P}} = a \sin\theta \dfrac{d\theta}{dt} \div a \dfrac{d\theta}{dt} = \sin\theta$.

We observe also that the motion may be divided into four equal portions or phases, and that the advance of Q from A to C has an exact counterpart on the return from B to C, and so for the other divisions; herein the motion possesses a simplicity which causes it to contrast in a marked degree with that discussed in Art. 7.

4. The motion of Q may be derived from that of P by the following arrangement:—

Let P represent a small pin set in a circular plate which is moveable about C as a centre of motion, and let the pin work in a groove, E F, whose direction is at right angles to that of the sliding bar, Q B. (Fig. 16.)

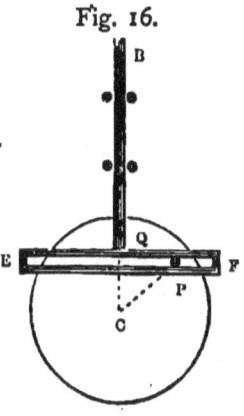

Fig. 16.

Of the two equivalents which combine to produce the circular motion of P, that which occurs in the direction F E is rendered inoperative, and the whole of the other equivalent is imparted to the sliding bar; in this way, then, the bar rises and falls as the disc rotates upon its centre.

5. We proceed now to analyse the conversion of circular into reciprocating motion by means of the *crank and connecting rod*.

A crank is merely a lever or bar moveable about a centre at one end, and capable of being turned round by a force applied at the other end; in this form it has been used from the earliest times as a handle to turn a wheel. When the crank is attached by a connecting rod to some reciprocating piece, it furnishes a combination which is extremely useful in machinery.

In the next chapter we shall see that the crank and connecting rod is one of the principal contrivances for converting reciprocating into circular motion; the student will understand that any such distinction as to the effect of the contrivance is one of classification only, regard being had to the direction in which the moving force travels. The arrangement is often used under both aspects in one and the same machine; as in a marine engine, where the piston in the steam-cylinder actuates the paddle-shaft by means of a crank and a connecting rod, and the motion is then carried on, by a crank forged upon the same shaft, to the bucket or piston of the air-pump.

It was in the year 1769 that James Watt published his invention of 'A method of Lessening the Consumption of Steam and Fuel in Fire-Engines,' the main feature of which was the condensation of the steam in a vessel distinct from the steam-cylinder. The steam-engine was at that time called a fire-engine, and was used exclusively in pumping water out of mines; the steam piston and the pump rods were suspended by chains from either end of a heavy beam centered upon an axis, the action of the steam caused a pull in one direction only, and the pump rods being raised by the agency of the steam were afterwards allowed to descend by their own weight.

In this shape the steam-engine was entirely unfitted for actuating machinery, and it was not until after the impulse given by Watt's invention was beginning to be felt that it became apparent that the expansive force of steam could be made available as a source of power in driving the machinery of mills.

While Watt was occupied with this great problem of the construction of double-acting engines, which eventually he fully solved, it happened that one James Pickard, of Birmingham, in the year 1780, took out a patent for a 'new invented method of applying steam or fire engines to the turning of wheels,' in which he proposed to connect the great working beam of the engine with a crank upon the shaft of a wheel by means of a spear or connecting rod, jointed at its extremities to the beam and crank respectively.

It is probable that Watt had foreseen this application of

The Crank and Connecting Rod.

the crank as early as the year 1778, and had intended to apply the combination as a means of carrying on the power from the end of the working beam to the fly-wheel; being forestalled, however, by the patentee, he did not dispute the invention, and contented himself with patenting certain other methods of obtaining a like result, among which will be found the sun and planet wheels described in a subsequent chapter.

This latter invention served his purpose until the patent for the crank had expired, and then it was that the more simple arrangement which we are now about to discuss came into general use.

6. The manner of employing the crank and connecting

Fig. 17.

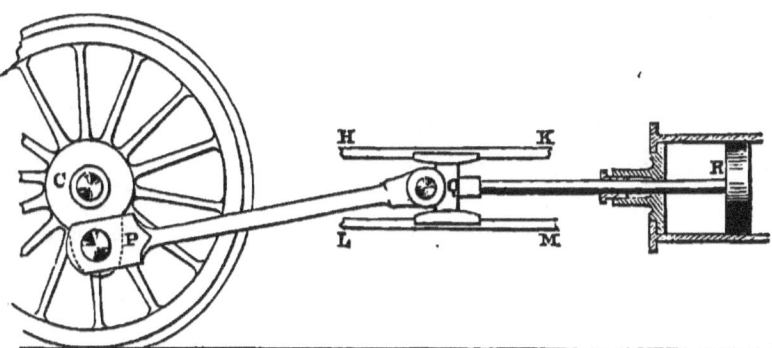

rod in the locomotive engine is shown in Fig. 17. The crank C P is made a part of the driving wheel of the engine, the connecting rod P Q is attached to the end of the piston rod Q R, and the end Q is constrained to move in a horizontal line by means of the guides H K, L M.

The contrivance here referred to belongs properly to the second chapter, as we are now engaged upon the conversion of circular into reciprocating motion.

We proceed to determine the relative positions of the crank and connecting rod during the motion.

7. Let C P represent an arm or *crank* centred at C, and connected by means of a link or *connecting rod*, P Q, with a point Q, which is constrained to move in the line C E D.

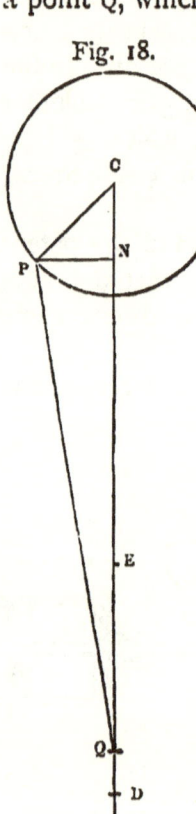

Fig. 18.

Draw $PN \perp^r CD$, and let $CP = a$,
$PQ = b$, $PCQ = C$, $PQC = Q$.
Then $CQ = CN + NQ$
$= a \cos C + b \cos Q$

Also $\dfrac{\sin Q}{\sin C} = \dfrac{a}{b}$ ∴ $\sin Q = \dfrac{a}{b} \sin C$

and $\cos Q = \sqrt{1 - \dfrac{a^2 \sin^2 C}{b^2}}$

∴ $CQ = a \cos C + \sqrt{b^2 - a^2 \sin^2 C}$

which gives the position of Q for any value of C, that is, for any given position of the crank C P.

Cor. 1. Let $C = 0°$ ∴ $CD = a + b$
$C = 180°$ ∴ $CE = -a + b$
whence $DE = CD - CE = 2a$.

The space D E is called the *throw* of the crank.

Cor. 2. If the position of Q be estimated by its distance from D, we have
$DQ = CD - CQ$
$= a + b - (a \cos C + b \cos Q)$
$= a(1 - \cos C) + b(1 - \cos Q)$.

But we have just shown that an expression such as $a(1 - \cos C)$ represents the resolution of circular into reciprocating motion, and we infer that the motion of the point Q is compounded of the resolved parts of two circular motions, one being that due to the motion of P in a circle round C, the other, that

Note.—Here also as in Art. 3, if $PCQ = \theta$, we have

vel. of $P = a \dfrac{d\theta}{dt}$, vel. of $Q = a \sin\theta \dfrac{d\theta}{dt} + \dfrac{a^2 \sin\theta \cos\theta}{\sqrt{b^2 - a^2 \sin^2\theta}} \dfrac{d\theta}{dt}$,

whence $\dfrac{\text{vel. of Q}}{\text{vel. of P}} = \sin\theta + \dfrac{a \sin\theta \cos\theta}{\sqrt{b^2 - a^2 \sin^2\theta}}$.

resulting from the motion of P in a circular arc through an angle Q, and produced by the swinging of the rod P Q about one end Q as it moves to and fro.

In other words, the connecting rod introduces an inequality, which prevents the motion of the point Q from retaining that evenness and regularity of change which is found in the motion of the point N; we now see by analysis that this inequality, whereby the motion of Q differs from that of N, is equal to $b(1-\cos Q)$.

Ex. Let C P = 10 inches, P Q = 5 feet, as in the case of the locomotive in Fig. 17; find the position of the piston when the crank is vertical.

Here $\sin Q = \dfrac{CP}{PQ} = \dfrac{10}{60} = \dfrac{1}{6}$

$\therefore \cos Q = \sqrt{1-\dfrac{1}{36}} = \sqrt{\dfrac{35}{36}} = \dfrac{1}{6}\sqrt{35}$

$\therefore DQ = a(1-\cos c) + b(1-\cos Q)$

$= 10 + 60\left(1-\dfrac{1}{6}\sqrt{35}\right)$

$= 10 + \cdot 84$ nearly,

or Q is nearly six-sevenths of an inch in advance of the centre of its path when the crank has made a quarter of a revolution from the line C D.

This inequality, arising from the shortness of the connecting rod, produces a sensible effect in the working of direct acting engines, and tends to make the mean pressure of the steam greater in the up stroke, where the piston is gaining in its motion, than in the down stroke, where it is lagging behind.

As the connecting rod is shortened the inequality increases, and the motion becomes more unequal.

Take a very extreme case, where $PQ = CP = a$,

$\therefore DQ = a(1-\cos c) + a(1-\cos Q)$.

Let now C = 60°, then Q = 60° also, because the triangle C P Q is now isosceles,

$$\therefore \cos C = \frac{1}{2} = \cos Q$$

$$\therefore DQ = 2a - a = a,$$

or Q has moved through half its path, while C P has turned through an angle of only 60°.

When $C = 90$, the point Q comes to C, and there the motion ends, for the crank C P can now go on rotating for ever without tending to move Q in any way.

Cor. 3. If the connecting rod could be prolonged until it became infinite, we should have P Q always parallel to itself, or $Q = 0$, and in that case the travel of Q would be represented by the equation $DQ = a(1 - \cos C)$.

A crank with a connecting rod of infinite length is an imaginary creation, but we shall presently see that an equivalent motion may be obtained in various ways, and we have already met with it in Art. 3.

8. The *eccentric circle* supplies a ready method of obtaining the motion given by a crank and link, and we proceed to examine it with the intention of ascertaining by what expedients we are enabled to vary the lengths of the particular crank and link which exist in every form of the arrangement.

And, first, we notice that the length of the crank is in every case equal to the distance between the centre of motion and the centre of the eccentric circle; it is, in fact, the line, C P, in each of the annexed drawings.

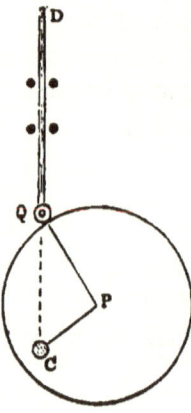

Fig. 19.

Let us consider the motion shown in Fig. 19, where a circular plate, moveable about a centre of motion at C, imparts an oscillatory movement to a bar, Q D, which is capable of sliding between guides in a vertical line, D Q, pointing towards C. Since P Q remains constant as the plate revolves, it is evident that Q moves up and down in the line C D, just as if it were actuated by the crank, C P, and the con-

The Eccentric Circle.

necting rod, P Q; the length of the connecting rod is in this case, therefore, equal to the radius of the rotating circle. It is obvious that an arrangement of this kind would be little used, by reason of the oblique thrust on the bar Q D.

9. A second form which, however, is of the greatest possible value, is deducible at once from that last examined.

Instead of allowing the end of the bar, Q D, to rest directly upon the circumference of the circle, suppose that bar to terminate in a half hoop which fits the circle, as shown in the drawing; let the rod point to the centre of the circle, and let one end, Q, be compelled to move in a line pointing to C. As the circle revolves it is evident that we have a crank, C P, just as before; but we have, in addition, a link which is now represented by P Q, and which may extend as far as we please beyond the limits of the circular plate.

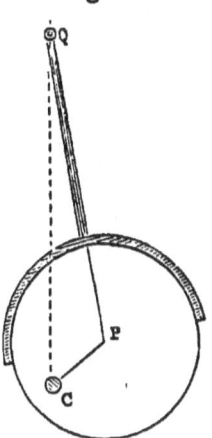

Fig. 20.

We thus obtain a combination which is sometimes described as a mechanical equivalent for the crank and connecting rod.

The form usually adopted in practice is derived at once from the arrangement just described. A circular plate is completely encircled by a hoop, to which a bar is attached: this bar always points to P, the centre of the plate, and its extremity drives a pin Q, which is constrained to move in the line C Q. (Fig. 21.)

The plate is moveable about a centre of motion at C, and we have already explained that P Q remains constant during each revolution of the plate, or that the resulting motion impressed upon Q is that due to a crank, C P, and a link, P Q.

As before, the *throw of the eccentric* is the same as that of the crank, viz., a space equal to the diameter of the circle whose radius is C P.

We should remark that P, the centre of the plate in question, may be brought as near as we please to C, the centre of the shaft, and that the throw of the eccentric may be reduced accordingly; but that we are limited in the other direction, for the shaft must be kept within the boundary of the plate, and the plate itself must not be inconveniently large; considerations which are sufficient to prevent our increasing C P in any great degree.

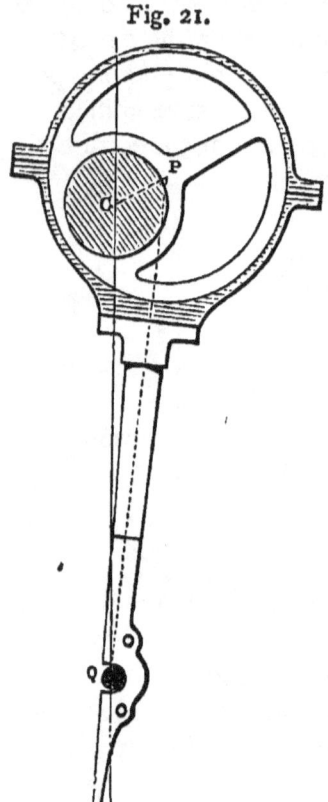

Fig. 21.

The eccentric circle may also be regarded as a simple form of *cam* (see Art. 18), but we have examined it here on account of its being identical in principle with the crank and connecting rod.

The object of the complete hoop is to drive Q in alternate directions. In some cases Q is brought back by a spring, and then only half the hoop is required; an instance occurs in modern forging machines, where the motion is very small and rapid.

On referring back to Fig. 18, it will be seen that the crank and connecting rod labour under the disadvantage of entailing a division of the shaft whenever it is required to place the crank anywhere except at one end, for the connecting rod is continually traversing over the centre of the shaft.

If, therefore, a crank is wanted in some intermediate portion of an axle or shaft, the axle must be *cranked* in the manner shown in Fig. 22, or be divided, and the two cranks or arms will be connected by a pin. These cranks and the pin are frequently forged in one solid mass upon the shaft, and shaped afterwards by the machinery of the workshop.

The great value of the eccentric arises from the circumstance that it enables us to derive the motion, which would be given by a crank, from any part of a shaft without the necessity of subdividing it; this is particularly noticeable in the mechanism of the steam-engine, where the crank of small throw which is required for moving the steam slide-valves, is almost invariably furnished by the aid of an eccentric keyed upon the main shaft.

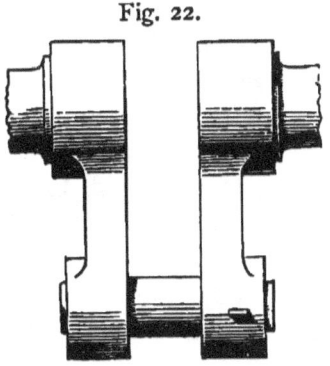

Fig. 22.

10. There is yet another method of arranging the eccentric circle, which gives us the combination of a *crank* with an *infinite link*.

We here intend, as explained in a former article, that the reciprocating piece shall be driven by a crank and connecting rod in such a manner that the connecting rod shall always remain parallel to itself, a result which could only happen in theory if the connecting rod were indefinitely lengthened.

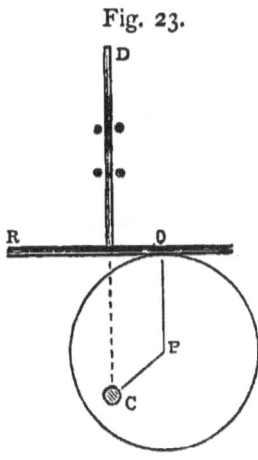

Fig. 23.

Suppose the roller at Q to be replaced by a cross bar Q R, standing at right angles to D C; as the circle revolves, it will cause the bar to reciprocate, C P will remain constant, and P Q will always be at right angles to R Q, and will therefore remain parallel to C D in all positions.

But this is the motion of a crank with an infinite link.

If C were in the circumference of the circle, the motion would be just the same, except that now the crank would be the radius of the eccentric circle. We shall presently notice a useful illustration of this particular case.

Take the following as an example of the movement under discussion.

In Mr. Anderson's machine for compressing elongated rifle bullets, there are punches fixed at the two ends of a strong massive rod, to which a reciprocating motion in a horizontal line is imparted, and a piece of lead is compressed into the required form at each end alternately.

The object of one part of the mechanism is to cause this rod to reciprocate, and the movement is obtained as follows.

Fig. 24.

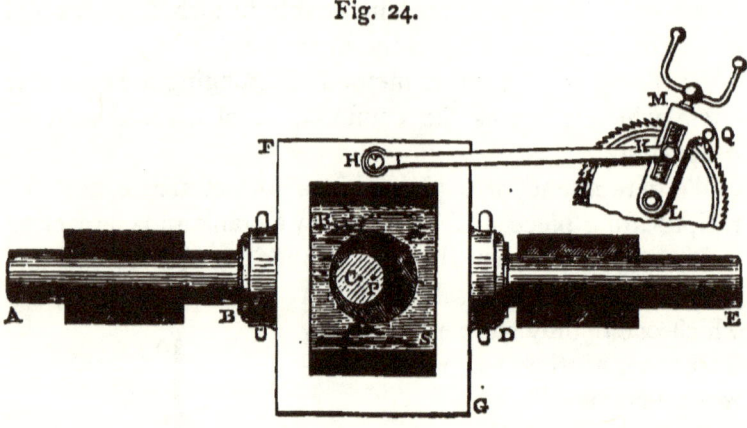

A small circle centred at C represents in section a shaft caused to rotate by the power of an engine; upon this shaft a short cylindrical block is forged so as to form part of it, and is afterwards accurately turned into the form of a circular cylinder, whose axis passes through P upon one side of the original axis.

A rectangular brass block R S, is bored out to fit the larger cylinder and slides in the rectangular frame F G, to which the cylindrical pieces A B, D E, which carry the punches, are

The Swash-Plate.

attached; the whole is put together in the manner shown, and it requires very little effort to understand that the rotation of the eccentric cylinder round an axis through C will impart to R S simultaneous movements in a horizontal and vertical direction, whereof one, viz. that in a direction perpendicular to A E, will be inoperative, and the other will be communicated directly to F G, and so to the rods carrying the punches. Thus A B and E D will be made to oscillate in the guides indicated in the sketch. In this case, then, the eccentric circle, whose centre is at P, is caused to rotate about the point C, and gives motion to the sides of the frame F G, just as it moved the bar Q R in the last article, and hence the resulting motion is that of a crank C P, with an infinite link.

The contrivance of the ratchet wheel at the right hand will be explained in the next chapter.

11. The crank with an infinite link also appears under the guise of a *swash-plate*.

Here a circular plate, E F (Fig. 25), is set obliquely upon an axis, A C, and by its rotation causes a sliding bar P Q, whose direction is parallel to A C, to oscillate continually with an up and down movement, the friction between the end of the bar and the plate being relieved by a small roller.

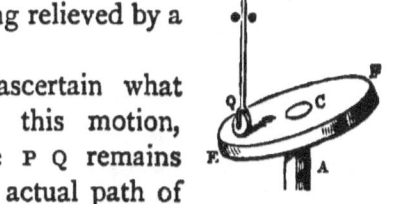

Fig. 25.

We must now try and ascertain what is the law which governs this motion, and we observe that since P Q remains always parallel to A C, the actual path of Q, as projected upon an imaginary plane through the lowest position of Q, and perpendicular to A C, will be a circle.

If this be so, it follows that the path of Q upon the plate itself will not be a circle, but an oval curve, and, as a matter of geometry, we can prove that the line C Q will vary in length as it rises or falls during the rotation of the plate, in the precise degree necessary for the description of the curve known as an ellipse.

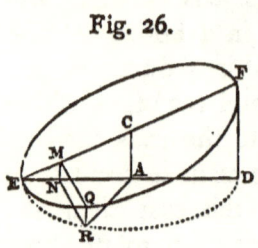

Fig. 26.

In Fig. 26 let E Q F represent the actual path of Q on the plate, and let the circle E R D be the projection of this path upon a plane $\perp^r$ to the axis A C.

Draw Q M $\perp^r$ to E F, Q R $\perp^r$ to the plane E R D, and R N $\perp^r$ to E D, which is the diameter of the circle E R D.

Join M N, and suppose the plate to rotate through an angle E A R = A, and thus to carry the roller at Q through a vertical space equal to R Q.

Then
$$R Q = M N = A C \times \frac{E N}{E A}$$
$$= A C \left(\frac{E A - A N}{E A} \right)$$
$$= A C \left(1 - \frac{A N}{A R} \right)$$
$$= A C (1 - \cos A);$$

or the motion is that of a crank A C with an infinite link.

This is a curious result; hitherto, in the motion of a crank with an infinite link, the reciprocation has always taken place in a plane perpendicular to the axis of rotation, but here we get the very same movement in a plane which contains the axis instead of being perpendicular to it.

12. It is sometimes required that the reciprocating motion shall be intermittent, or have intervals of rest.

This motion may be provided for by placing a loop at the point where the eccentric bar engages the pin. It is evident that the pin will only move when one end of the loop takes it up; but in doing this a blow is struck, which it may be well to avoid, and hence an intermittent motion has been obtained in a much better manner by a movement adapted for working the slide-valves of a steam-engine.

Valve-Motion.

We can readily see that if any portion of the plate in Art. 10 be shaped in the form of a circle round C, such portion will have no power of moving the sliding bar.

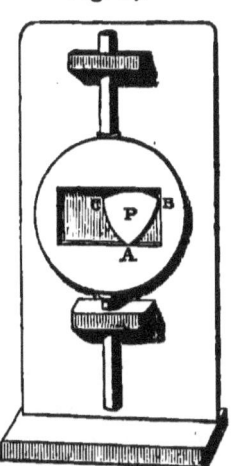

Fig. 27.

Let the pin P assume the form of a circular equilateral triangle, C A B, formed by three circular arcs, whose centres are in A, B, and C respectively, and let it be embraced by a rectangular frame attached to a sliding rod.

As C A B revolves round the point C, the portion C B will raise the plate; the point B will next come into action, and will raise the plate still higher; the upper edge of the groove will then continue for a time upon the curved surface A B, which is a circular arc described about C as a centre, and here the motion will cease; the plate will next begin to fall, will descend as it rose, an interval of rest will succeed, and thus we shall produce an intermittent movement, which may be analysed as follows:—

Suppose the circle described by B to be divided into six equal parts, at the points numbered 1, 2, 3, 4, 5, 6.

Fig. 28.

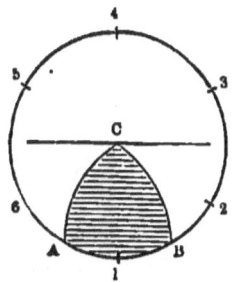 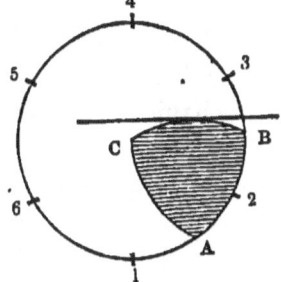

As B moves from 1 to 2, the frame remains at rest; from 2 to 3 the arc C B drives the frame, the centre of motion of the eccentric circle being now a point in its circumference, and the horizontal bar is driven as it would be by a crank

Fig. 28 (a).

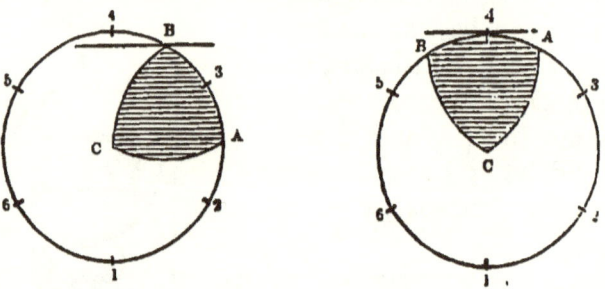

C B with an infinite link (*see Art.* 10); from 3 to 4 the point B drives, and the motion is again that of a crank C B with an infinite link (*see Art.* 4); i.e. the motion from 3 to 4 is the same as that from 2 to 3, except that it is decreasing in velocity instead of increasing.

From 4 to 5 there is rest, then an increase of motion from 5 to 6, and finally a decrease to zero as B passes through the arc 6 to 1 and completes an entire revolution.

13. We have hitherto confined our attention to one simple example in the geometry of motion: we shall now extend our view on the subject, and shall consider the driver to possess circular motion whenever it rotates continuously upon a fixed centre or axis; and in order to generalise still further, we shall suppose the reciprocating motion to be either rectilinear or circular; we shall in this manner be enabled to bring under one point of view a great variety of useful mechanical contrivances.

14. Circular may be converted into reciprocating motion by the aid of *escapements*.

An *escapement* consists of a wheel fitted with teeth which are made to act upon two distinct pieces or *pallets* attached to a reciprocating frame, and it is arranged that when one tooth escapes, or ceases to drive a pallet, the other shall commence its action.

One of the most simple forms is the following:—

A sliding frame, A B, is furnished with two projecting pieces at C and D, and within it is centred a wheel possessing three teeth, P, Q, and R, which tends always to turn in the direction indicated by the arrow.

The upper tooth, P, is represented as pressing upon the projection D, and driving the frame to the right hand: when

Fig. 29.

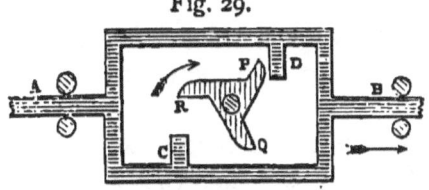

the tooth P escapes, the action of Q commences upon the other side of the frame, and the projection C is driven to the left hand. Thus the rotation of the wheel causes a reciprocating movement in the sliding piece, A B.

It is clear that the wheel must have 1, 3, 5, or some odd number of teeth upon its circumference.

15. The *Crown Wheel* escapement was invented for the earliest clock of which we possess any record.

The form of the wheel is that of a circular band, with large saw-shaped teeth cut upon one edge; the vibrating axis A B (Fig. 30), carries two flat pieces of steel, a, b, called pallets, which project from the axis in directions at right angles to each other, and engage alternately with teeth upon the opposite sides of the wheel. Suppose the wheel to turn in the direction towards which the teeth incline, and let one of its teeth encounter the pallet b and push it out of the way; as soon as b escapes, a tooth on the

Fig. 30.

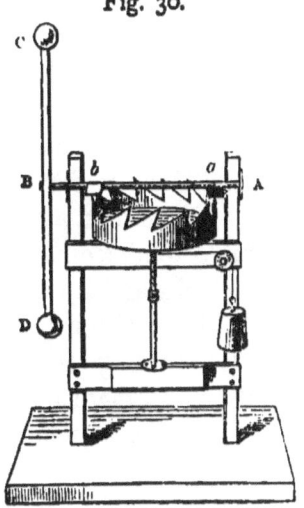

opposite side meets the pallet *a* and tends to bring the axis A B back again : thus a reciprocating action is set up, which will be very rapid unless A B is provided with a heavy arm, C D, at right angles to itself. Such an arm possesses *inertia*, so that its motion cannot be suddenly checked and reversed, and a recoil action is set up which materially subtracts from the utility of this contrivance. For it will be seen that the vibration of C D cannot be made to cease suddenly, and that the wheel must of necessity give way and recoil at the first instant of each engagement between a tooth and its corresponding pallet.

The more heavily C D is loaded at a distance from the axis the more slowly will the escapement work, and the greater will be the amount of the recoil.

Here we have an invention which has done good service to mankind. It was used in the first clock which was ever made, and dealt out time through the *step by step* movement of the wheel with pointed teeth. This wheel, urged on by a weight, and hampered always by the vibrating bar whose pallets were perpetually getting into the way of its teeth, moved round with a slow, intermittent, and step by step movement, checked and advancing alternately, but solving for mankind, in a clumsy though tolerably accurate manner, the great problem of the mechanical measurement of time, and giving birth, by the idea suggested, to those marvellous pieces of mechanism which have finally resulted in the modern astronomical clock and the chronometer.

Being, no doubt, a coarse and imperfect arrangement, it has gradually sunk from one level to another; it has disappeared from clocks and from watches also, and is now seldom to be met with except in the homely contrivance of the kitchen-jack for roasting meat.

16. A great advance in the progress of mechanism was made by Dr. Hooke, the contemporary of Newton, who devised the so-called *Anchor Escapement*.

Here a wheel centred at E is provided with a number of

teeth, and tends always to turn in the direction indicated by the arrow.

A portion of this wheel is embraced by an anchor, A C B, centred at C, the extreme ends of which are formed into pallets, A *m* and B *n*: these pallets may be flat or slightly convex, but they are subject to the condition that the perpendicular to A *m* shall pass above C, and the perpendicular to B *n* shall pass between C and E. The point of a tooth is represented as having escaped

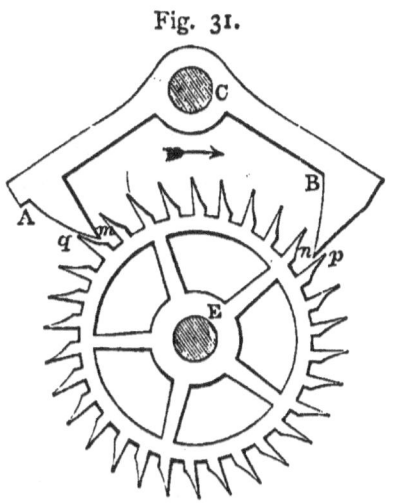

Fig. 31.

from the pallet B *n* after driving the anchor to the right hand; and the point *q*, by pressing against A *m*, is supposed to have already pushed the anchor a little to the left hand, and thus the wheel can only proceed by causing a vibratory motion in the anchor, A C B.

The rapidity of this vibration may be very great, or it may be reduced by connecting the anchor with a much heavier body, such as the pendulum of a clock.

There is the same recoil experienced upon each swing of the pendulum as that which we noticed in the last article, and the contrivance is commonly known as the *Recoil Escapement*.

The exact character of the action which takes place between the pendulum of a clock and the scape wheel has been the subject of a long and interesting mathematical investigation: it is foreign to our purpose to discuss it at present, but we may state in general terms the nature of the problem.

The going part of a clock consists of a train of wheels

tending to move under the action of a weight or spring: if the last wheel of the train were left to itself, it would spin round with great velocity, and we should fail in obtaining any measure of time.

The escapement is one part of a contrivance for regulating the velocity of the train of wheels, but the escapement alone is not sufficient; we require further a vibrating body possessing *inertia*, the motion of which cannot be suddenly stopped or reversed.

Such a body is found in the pendulum, and a very intricate mutual action exists between the pendulum and the scape wheel. The function of the pendulum is to regulate and determine the periods and amount of onward motion in the scape wheel, whereas the office of the wheel is to impart such an impulse to the pendulum at each period of this onward movement as may serve to maintain its swing unimpaired, and may cause it to move with the same mathematical precision which would characterise the vibrations of a body swinging *in vacuo*, and uninfluenced by any disturbing causes.

17. The teeth in the wheel are sometimes replaced by pins, in which case the form of the anchor may be so altered that the action shall take place upon one side of the wheel, as shown in Fig. 32.

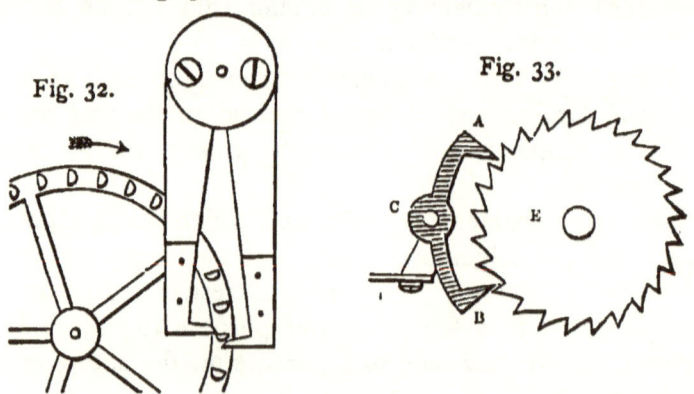

Fig. 32.

Fig. 33.

In a printing telegraph instrument the recoil escapement has been employed to control the rapidity of motion in a train of wheels, and the number of vibrations of the anchor are appreciated by listening to the musical note which it imparts to a vibrating spring.

The anchor A C B (Fig. 33) is centred at C, and vibrates rapidly as the scape wheel E revolves; a strip of metal F carries on the oscillation to a steel spring which gives the note, and the velocity of the train can be regulated by an adjustable weight attached to the spring.

Again, the same escapement forms part of the mechanism of an *Alarum clock*, where a hammer is attached by a bar to the anchor, and blows are struck upon the bell of the clock in rapid succession as the scape wheel runs round.

18. Circular may be converted into reciprocating motion by the aid of *cams*.

The term 'cam' is applied to a curved plate or groove which communicates motion to another piece by action of its curved edge.

Such a plate is shown in Fig. 34, and, as an illustration, we shall suppose that the portions ab, ca are any given curves, and that bc is a portion of a circle described about the centre of motion.

It is easy to understand, that as the cam rotates in the direction of the arrow, the roller P at the end of the lever A P will be raised gradually by the curved portion ab, will be held at rest while bc passes underneath it, and, finally, will be allowed to fall by the action of ca.

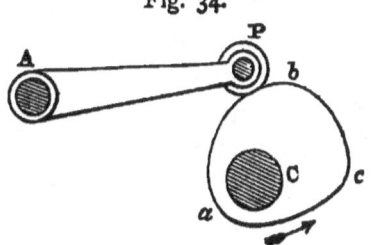

Fig. 34.

In this way a cam may be made to impart any required motion, and may reproduce in machinery those delicate and rapid movements which would otherwise demand the highest effort of skill from a practised workman.

19. The circular motion being uniform, the reciprocating piece may also move uniformly, or its velocity may be varied at pleasure.

1. Suppose that the reciprocating piece is a sliding bar, whose direction passes *through the centre of motion* of the cam plate; take C as this centre, let B P represent the sliding bar, and let A be the commencement of the curve of the cam-plate.

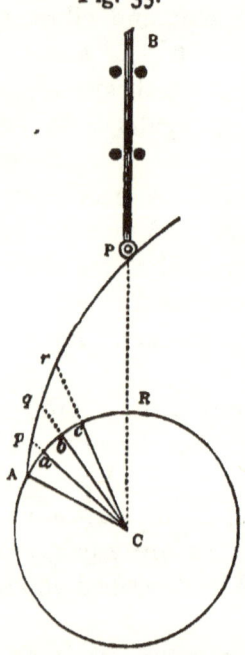

Fig. 35.

The curve A P may be set out in the following manner.

With centre C and radius C A describe a circle, and let B P produced meet its circumference in the point R.

Divide A R into a number of equal arcs A *a*, *a b*, *b c*, &c.

Join C *a*, C *b*, C *c*, &c., and produce them to *p*, *q*, *r*, &c., making *a p*, *b q*, *c r*, &c., respectively equal to the desired movements of B P in the corresponding positions of the cam-plate; the curve A *p q r*...P will represent the curve required.

This curve will often present in practice a very irregular shape, but in the particular case where the motion of P B is required to be uniform, it assumes a regular and well-known form (*see note*).*

2. We will next examine the case where the centre of motion of the cam-plate lies upon one side of the direction of the sliding bar, and we shall find that the method of setting out the curve changes accordingly.

Suppose that the direction of B P passes upon one side of the centre of motion C, draw C R perpendicular to B P pro-

---

* Let C A = a, C P = r, P C A = θ, and let B P move in such a manner that the linear velocity of P shall be m times that of the point A, in other words, let R P = m . R A.

Now R P = $r - a$, and R A = $a\theta$.

$$\therefore r - a = m . a\theta,$$

which is the equation to the spiral of Archimedes.

duced, describe a circle of radius C R, and conceive the motion to begin when A coincides with R (Fig. 36).

As a matter of theory such an extreme case is possible, and we will imagine it to exist in order to obtain the equation which represents the complete curve. Practically, the cam would be more effective in straining the bar than in moving it when the point P was near to the point R.

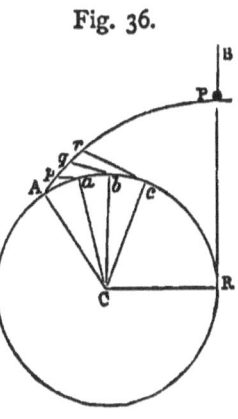

Fig. 36.

Divide A R into the equal intervals A a, $a\, b$, $b\, c$, &c., but now draw $a\, p$, $b\, q$, $c\, r$, &c., tangents to the circle, and equal in length respectively to the desired movements of B P during the corresponding periods of motion of the cam-plate.

The curve A $p\, q\, r$ ... P will be that required, and the analytical representation of it will be given in the note.*

* Let C P $= r$, C A $= a$, A C P $= \theta$, P C R $= \phi$. The curve A P is now of such a character that the linear velocity of A shall be m times that of P, or, in other words, R A $= m$. R P.

But R A $= a(\theta + \phi)$, R P $= a \tan \phi$,

$\therefore a(\theta + \phi) = m a . \tan \phi.$

Now $\cos \phi = \dfrac{a}{r}$, and $\tan \phi = \sqrt{\dfrac{1 - \cos^2 \phi}{\cos^2 \phi}}$

$\therefore \tan \phi = \sqrt{\dfrac{r^2}{a^2} - 1}$

whence $\theta + \cos^{-1} \dfrac{a}{r} = m \sqrt{\dfrac{r^2}{a^2} - 1}.$

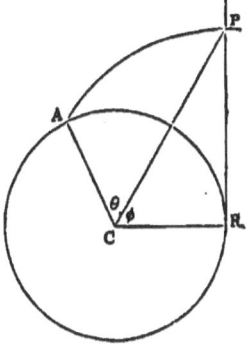

Fig. 37.

Cor. Let $m = 1$, or R A $=$ R P, which would happen if A P were a stretched string unwound from the circle; the curve traced out by the end P of the string becomes in this case a well-known curve called the involute of the circle, and our equation takes the form

$$\theta + \cos^{-1} \dfrac{a}{r} = \sqrt{\dfrac{r^2}{a^2} - 1},$$

which is the equation to the involute of a circle.

42   *Elements of Mechanism.*

20. In order to illustrate in a lecture the power of cams to produce any required movement, the late Professor Cowper arranged a model which would write the letters R I, selected probably as a compliment to the Royal Institution.

The principle of this combination of cams will be readily

Fig. 38.

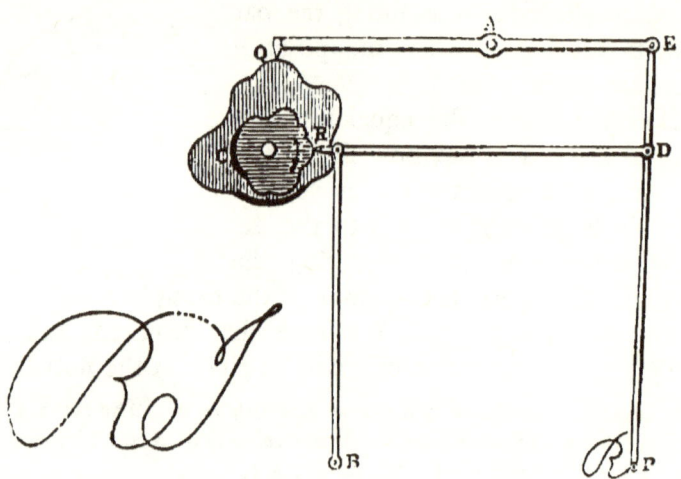

understood if we remember that the successive movements of a point in directions parallel to two perpendicular lines will suffice to enable the point to take up any position in the plane of the two given lines.

The bars shown in the drawing have fixed centres at A and B, and it is apparent that if we were to remove the cam and fasten the joint at R to the plane, we should be able to give P a vertical movement by the swing of the arms A E and R D. In the same way, if we fastened E to the plane and liberated R, the arm E P could swing about E, and P would then describe a small circular arc which would closely approach to a horizontal line.

Connect now the bars with the cam as in the sketch, and press the pointers at Q and R against the curves of the respective cams. Let these cams revolve slowly about a centre C (marked as a round spot in the smaller cam-plate), in the direction shown by the arrow, and the required letters will be traced out by the pencil at P.

In the figure the letter R has just been completed, and the pencil is about to trace out the lower tail of the letter I.

The two darkened lines in the cams are arcs of circles about the centre of motion where nothing is being done, the pencil remaining at rest while the cam rotates through a small angle.

This example shows us that a combination of two cam-plates actuating a simple framework of levers will give the command of any movement in a plane perpendicular to the axis of rotation: we shall presently see how to obtain a motion parallel to the same axis, and thus we can secure any required movement in space.

21. The heart wheel has been much used in machinery, and is formed by the union of two similar and equal cams of the character discussed in the first part of Art. 19.

A curved plate, C, shaped like a heart, actuates a roller, P, which is placed at the end of a sliding bar, or which may be attached to a lever P A B, centred at some point A, and connected by a rod B D to the reciprocating piece. The peculiar form of the cam allows it to perform complete revolutions, and to cause an alternate ascent or descent of the roller P with a velocity which may be made quite uniform.

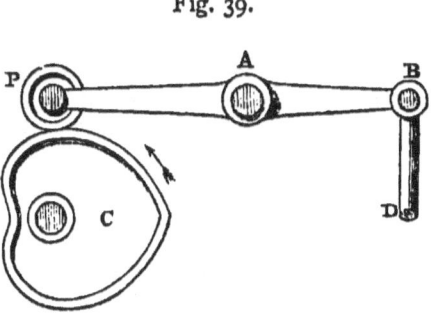

Fig. 39.

Since a cam of this kind will only drive in one direction, the follower must be pressed against the curve by the reverse action of a weight or spring.

22. The lever punching machine is worked by a cam resembling that which we give as an example by Mr. Fletcher of Manchester. The cam is here shown attached to the axis of the driving wheel, and the lever, which carries the punch in a slide connected with its shorter arm, is centred on the pin at E.

Fig. 40.

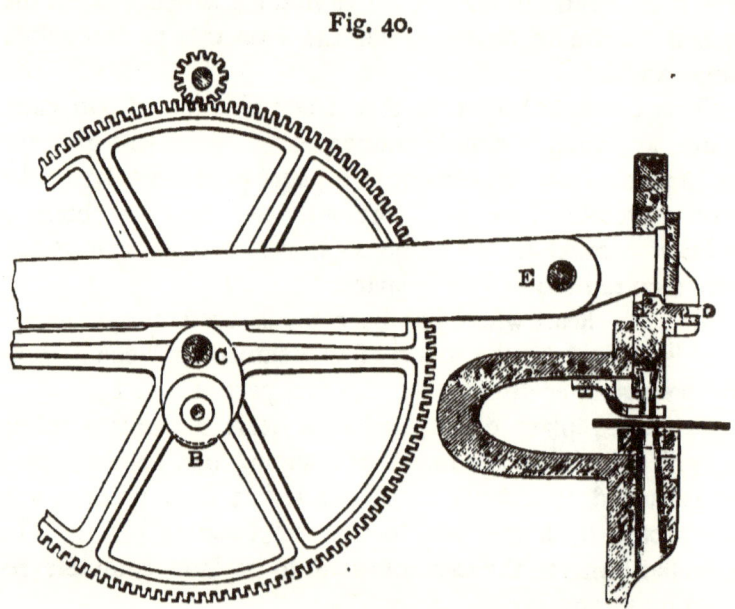

The curve of the cam is adapted to raising the longer arm of the lever bar in $\frac{1}{4}$ of a revolution of the driving shaft, it allows the lever to fall in the next $\frac{1}{4}$ of a revolution, and finally leaves the punch raised, as shown in the sketch, during the remaining $\frac{1}{2}$ of a revolution, thereby giving the workman an interval of time for adjusting the plate of iron before the next hole is punched.

That this action occurs will be quite evident upon inspecting the form of the cam, and it will also be seen that the cam is provided with a circular roller B, which determines the form of the driving surface while the work is being done, and which is merely an arrangement for lessening the friction just at the time when the greatest pressure is being exerted.

23. Hitherto we have considered the cam to be a plane curve or groove; but there is no such restriction as to its form in practice. Let us examine the following very simple case, as well as the extension of which it admits.

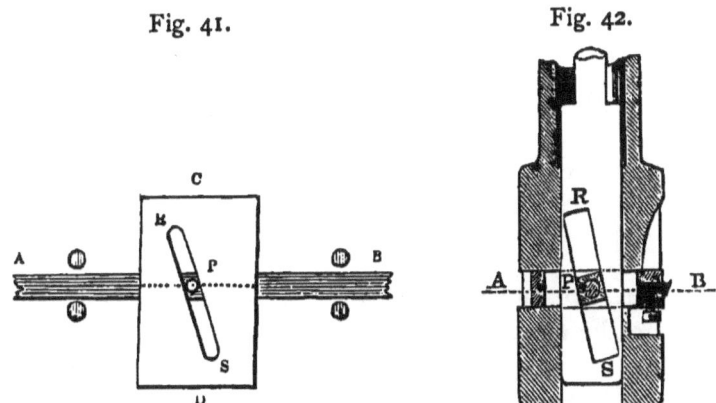

Fig. 41.   Fig. 42.

C D is a rectangle with a slit R S cut through it obliquely, a pin P fixed to the sliding bar A B works in the slit. If the rectangle C D be moved in the direction R S, it will impart no motion to the bar A B; but if it be moved in any other direction, the pin P will be pushed to the right or left, and a longitudinal movement will be communicated to the bar A B.

The contrivance here sketched is of frequent use in some form or other, and we may point out its application in the rifling bars used at Woolwich in the manufacture of rifled guns. In work of this kind, where the greatest accuracy is demanded, the bore of the gun acts as a guide to the head of the rifling bar, and the cutter does its work while the bar is being twisted and pulled out of the gun. It is essential, therefore, to keep the cutter within the head while the bar is being inserted preparatory to the removal of a strip of the metal, and to bring it out again at the end of the stroke.

In order to arrive at this result the bar is made hollow,

and the tool-holder in the rifling head, shown in Fig. 42, is made to move in and out laterally by means of a pin P working in an inclined slot R S, in the internal feed rod. As the feed rod is pushed through a small definite space in either direction around the axis of the bar, the cutter will also move in or out in the direction of the dotted line A B.

Fig. 43.

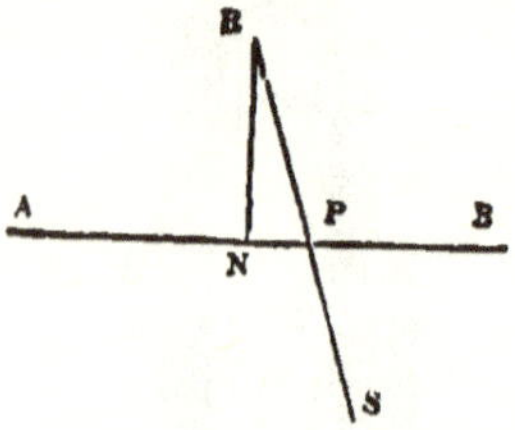

In discussing this motion there are two cases to examine.

1. Suppose that C D is moved at right angles to A B.

Draw R N $\perp^r$ A B.

$$\frac{\text{travel of C D}}{\text{travel of A B}} = \frac{R N}{P N}$$

$$= \tan R P N.$$

2. Let C D move in a direction inclined at any given angle to the direction of the groove R S.

Fig. 44.

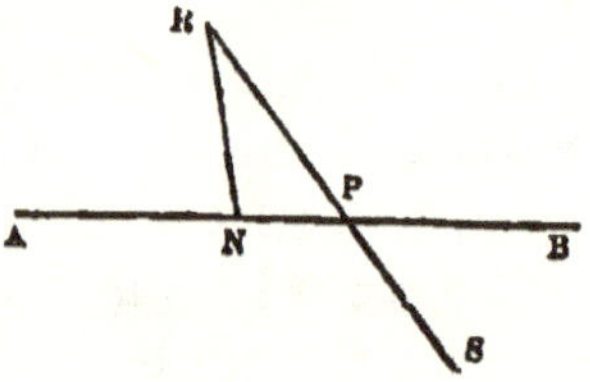

Draw R N in this direction, and we have

$$\frac{\text{travel of C D}}{\text{travel of A B}} = \frac{R N}{N P}$$

$$= \frac{\sin R P N}{\sin N R P}$$

In other words, the *velocity ratio* of C D to A B is expressed by the fraction $\frac{\sin R P N}{\sin N R P}$, and takes the form tan RPN when the angles at R and P make up a right angle.

24. Next let C D be wrapped round a cylinder; it will form a screw-thread, and the revolution of the cylinder upon its axis will be equivalent to a motion of the rectangle at right angles to the bar, in the manner shown in the preceding article; we shal have, therefore,

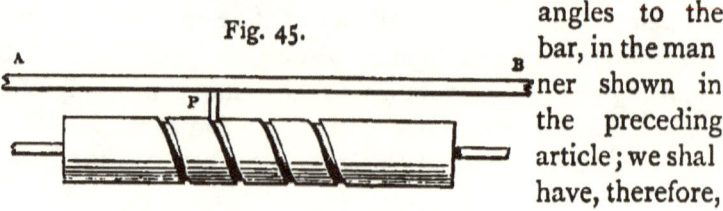

Fig. 45.

by the arrangement in the figure, a continuous uniform rectilinear motion of the bar A B during the revolution of the cylinder upon which the screw-thread is traced.

If the pitch of the screw be constant, the motion of P B will be uniform, and any change of velocity may be introduced by a proper variation in the direction of the screw-thread.

If the screw be changed into a circular ring, A B will not move at all.

It is, then, a matter of indifference whether the cam be a groove traced upon a flat plate or a spiral helix running round a cylinder. In the first case motion ensues when the groove departs from the circular form, and the distance from the centre varies; in the second case motion ensues the moment the groove begins to deviate from the form of a ring, whose plane is perpendicular to the axis.

As an illustration of a cam of the latter character, we may refer to the diagram, which shows a form very much used where a small motion of a lever is required; the lever A C B is centred upon the point C, and will commence to move as soon as the pin at the end A reaches that portion of the ring which departs from the circular form.

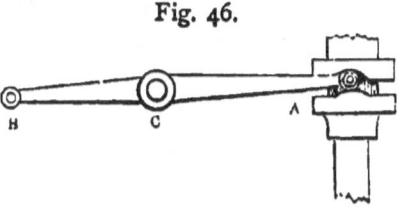

Fig. 46.

Note.—This kind of cam has the property of giving a motion parallel to the axis upon which it is shaped.

25. We subjoin an example, devised some twenty years ago, in which a reciprocating movement is imparted to a frisket frame in printing machinery, and it will be presently seen that the required result can be obtained in a much more simple manner.

The use of a cam plate allows of an interval of rest at each end of the motion, and enables the printer to obtain an impression, and to place a fresh sheet of paper upon the form.

Elements of Mechanism.

Here A H is the reciprocating frame attached to the combination of levers G F E D C B by the link A B. (Fig. 47.)

Fig. 47.

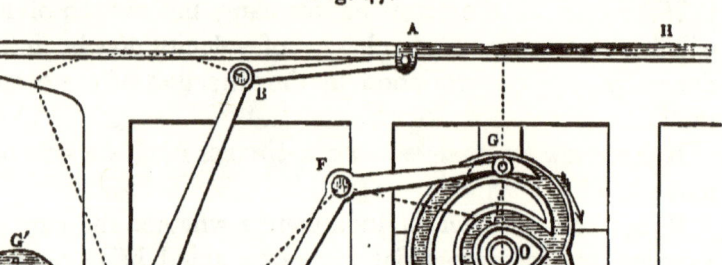

At the end of the lever, F G, is a sliding pin which travels along the grooves in the flat plate centred at O, and determines, by its position, the angular motion of the levers about the fixed centres at F and C.

Where the groove is circular, which occurs in those portions which are to the left hand of the vertical dotted line, the levers remain at rest, and they change into the position shown by the dotted lines when the sliding pin passes from the outer to the inner channel. The pin is elongated in form, as shown at G', and is thus capable of passing across the intersections of the groove.

Precisely the same character of movement may be obtained by the aid of a helical groove traced upon a revolving drum; the intervals of rest occur when the groove assumes the form of a flat ring, whose plane is perpendicular to the axis of the drum.

Fig. 48.

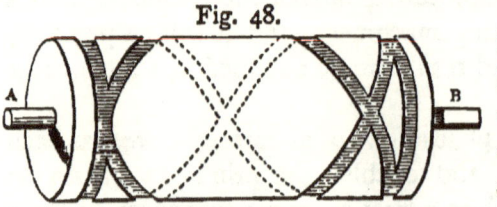

A right and left-handed screw-thread is traced upon the worm barrel, A B, which revolves in one uniform direction; a pin attached to the table of a printing machine follows the path of the groove upon the barrel, and its form is elongated so as to enable it to pass in the right direction at the points where the grooves intersect.

The interval of rest commences with the entry of the pin into the flat ring at either end of the barrel, and may be made to occupy the whole or any part of a revolution of A B, according as the grooves enter and leave the ring at the same or different points.

This construction dispenses with the complicated system of levers, which constitutes such a serious defect in the other arrangement.

Mr. Napier has patented an invention which causes the interval of 'rest' to endure beyond the period of one revolution of the barrel. (Fig. 49.)

At the entrance to the circular portion of the groove a moveable switch is placed, and it is provided that the switch shall be capable of twisting a little in either direction upon its point of support, and also that the pin upon which the switch rests shall admit of a small longitudinal movement parallel to the axis of the barrel, the pin itself being urged constantly to the right hand by the action of a spring.

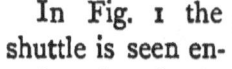

Fig. 49.

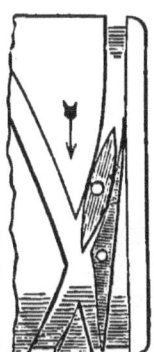
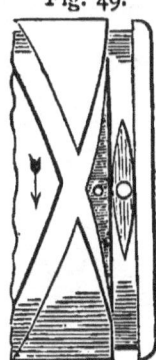
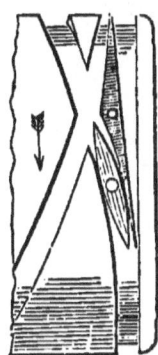

Fig. 1.   Fig. 2.   Fig. 3.

In Fig. 1 the shuttle is seen entering the circular portion of the groove, and twisting the switch into a position which will allow the shuttle to meet it again, as in Fig. 2, and to make a second journey round the circular ring.

The spring which presses the point of support of the switch to the right hand will now cause it to twist by means of the reaction which the passing pin affords, and the consequence will be, that the switch will be left in the position shown in Fig. 3, and will guide the shuttle into the helical portion of the groove. Thus the period of rest will be that due to about one and two-thirds of a revolution of the barrel.

26. Cams are employed when it is required to effect a movement with extreme precision. Thus in the machine of Mr. Applegath for printing newspapers, the sheet of paper starts upon its journey to meet the type at a particular instant of time; an error of one-twelfth of a second would cause the impression to deviate half a foot from its correct position, and would throw two columns of letter-press off the sheet of paper. The accuracy with which the sheet is delivered is therefore very remarkable, and is insured by the assistance of the cam represented in the diagram. (Fig. 50.)

As C revolves, the roller at B drops into the hollow of the plate, thereby determining the fall of the lever A B, and by it the fall also of another roller which starts the paper upon its course to the printing cylinder.

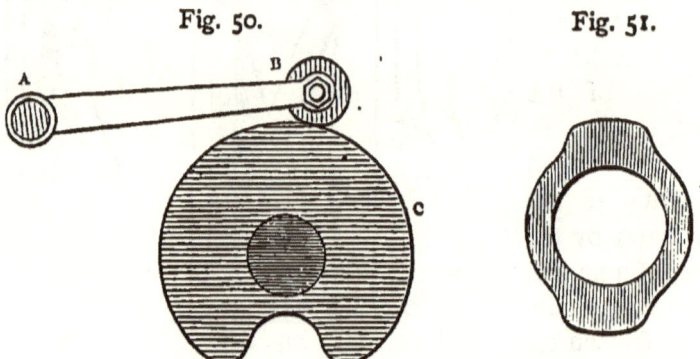

Fig. 50. Fig. 51.

Cams will give a rapid motion, with intervals of rest.

In the expansive working of a steam-engine there are two positions of rest for the valve which regulates the admission of steam into the slide-case, and since it is desirable to move the valve as rapidly as possible from one position to the other, we derive the form of an *expansion cam*. (Fig. 51.)

Cams. 51

27. Where the cam plate is required to effect more than one double oscillation of the sliding bar during each revolution, its edge must be formed into a corresponding number of waves.

Fig. 52.

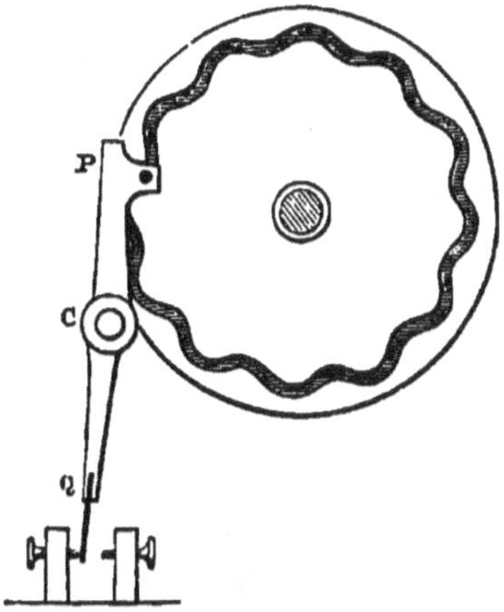

There is an example in telegraph commutators, the interruptions of the current being caused by the vibrations of a lever, P C Q, centred at c, and whose angular position is determined by a pin travelling in the groove.

As the wheel revolves, it can impress any given number of double oscillations upon the lever.

28. In the striking part of a large clock the hammer may be raised by a cam, and may then be suffered to fall abruptly.

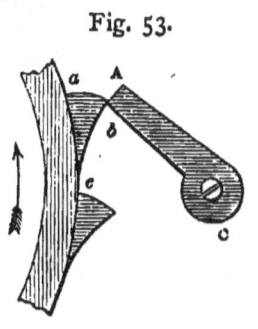

Fig. 53.

The figure represents the cam devised for the Westminster clock: the hammer rises and falls with the lever, A C, and the cam is so formed that its action commences at the extremity of the lever, and never departs sensibly from the same point; the cam, $a b$, is a circle whose centre is at the point of intersection of the tangents to the rim of the wheel at the points a and c.

29. We remark, in conclusion, that when the mechanic causes the moving body to be influenced by a pin which exactly fits the groove along which it travels, it is obvious that the moving body will take the exact position determined by the pin; on the other hand, where the cam is merely a curved plate pushing a body before it, there is no certainty that this body will return unless it be brought back by a weight or spring. Hence it arises that double cams have sometimes been employed in machinery, and we take the next example from an early form of power-loom.

A B is the treadle, E and F are the cam-wheels or tappets, which revolve in the directions shown by the arrows, and in

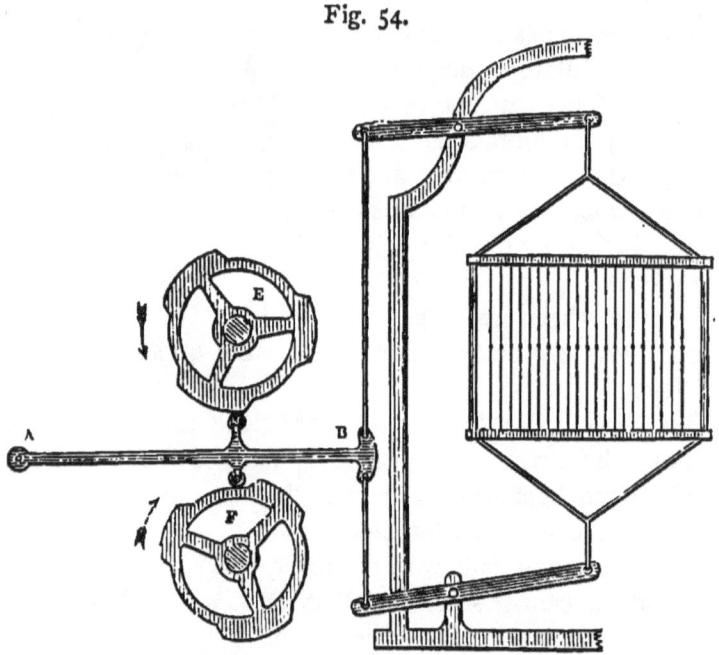

Fig. 54.

such relative positions that the projections and hollows are always exactly opposite to each other. As the cams rotate, the treadle, A B, is alternately elevated and depressed, and the threads of the warp are opened so as to permit the throw of the shuttle during the operation of weaving the fabric.

30. In the transfer of force by machinery, the moving power is carried from one piece of shafting to another, throughout the whole length and breadth of the factory; it passes from point to point, enters each separate machine, and gives movement to all the several parts which may be prepared for its reception.

Now it must be remembered, that the engine is never reversed, and that the power continues to flow onward in one uniform direction.

Take the case of a machine for planing iron: here the principal movement is that of a heavy table sliding forwards and backwards, and carrying the piece of metal which is the subject of the operation.

There are two methods of obtaining the desired result: the power may be poured, as it were, into the machine by a stream running always in one direction, and the reciprocation may be provided for by the construction of the internal parts, or the flow of the stream may be reversed by some intermediate arrangement external to the machine itself.

31. The former method is that usually adopted, and we shall now examine those machines where the reciprocation depends upon the internal construction of the moving parts.

And, first, we shall discuss a very simple and useful reversing motion which is obtained by a combination of two or three spur wheels, and which depends upon an obvious fact.

Fig. 55.

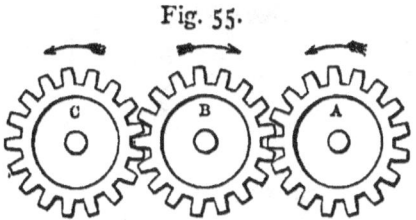

Let A, B, C, represent three spur wheels in gear; it will be seen that A and B turn in opposite directions, while A and C turn in the same direction. If then we connect two parallel axes by a combination of two and three spur wheels alternately, and properly arrange our driving pulleys, so that the power shall travel first through one combination and then through the other, we shall have a movement which has been adopted by Collier in his planing machines, and which has been subsequently much used by other makers.

The power is now derived from the shafting by means of a

band passing over a drum on the main shaft and over one of the three pulleys, E, I, F, at the entrance into the machine.

Of these pulleys E is keyed to the shaft, I rides loose upon it, while F is attached to a pipe or hollow shaft, through which the shaft connecting E with A' passes, and which terminates in the driving wheel A.

Fig. 56.

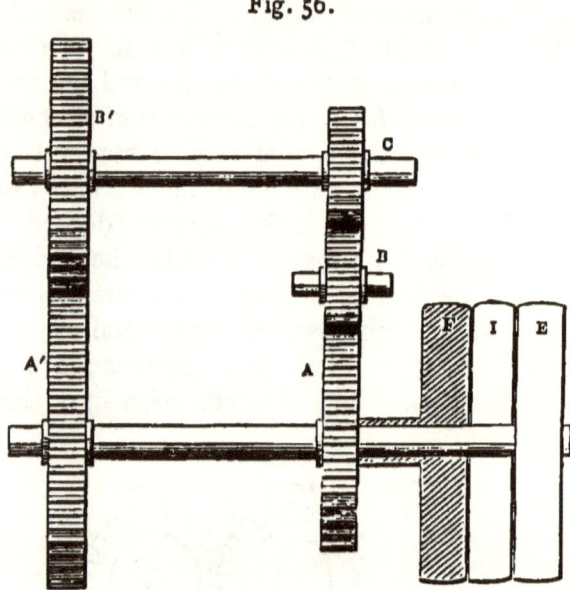

There is also a second shaft B' C, which carries the toothed wheels B' and C.

B is an intermediate wheel riding upon a separate stud.

When the band drives the pulley E, it is clear that A' and B' turn in opposite directions; whereas the motion is reversed when the band is shifted to F, for in that case A and C turn in the same direction. When the driving band is placed upon I, the machine remains at rest.

The rotation of B' C may be made much more rapid in one direction than in the other, and the construction is therefore particularly useful in machinery for cutting metals.

The slow movement occurs while the cutting tool is

Quick Return Motion.

removing a slip of metal, and the return brings the table rapidly back into the position suitable for a new cut.

32. This combination, slightly modified, is adopted generally in planing machines, and is valuable by reason of the uniformity of the movement, the rate of advance of the table being perfectly constant.

It also possesses the important advantage of saving in a great degree any useless expenditure of work, by causing the table to traverse with a *quick return* movement when the cutter is not in action.

We give so much of the machine as will explain the method of reversing the motion of the table. When the strap is upon the pulley F, the wheel A turns in one direction.

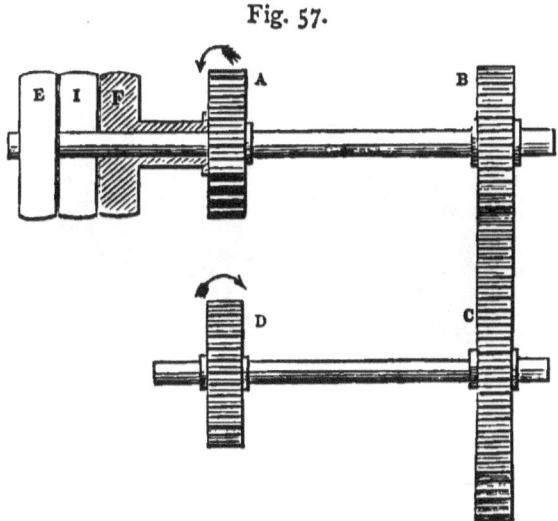

Fig. 57.

When the strap is upon the pulley E, the motion passes to B, which turns with E, and thus the axis, C D, is made to revolve in the opposite direction with a reduced velocity.

The wheels A and D both engage with another wheel not shown in the drawing, which actuates the table, and the reversal takes place when the moving power is transferred from the wheel A to D.

33. There is an instance of the application of this contrivance in a machine arranged for the purpose of cutting a screw-thread in the interior of the breech of an Armstrong gun.

In this case the driving pulleys are placed between the wheels A and A', and are formed in such a manner that the pulley F and the wheel A make one piece, and ride loose upon the shaft H K, as do also, in their turn, the pulley E and the wheel A'; the wheel M is keyed to H K, so as to rotate with it, and is further attached by a coupling to the muzzle of the gun which is to be operated upon.

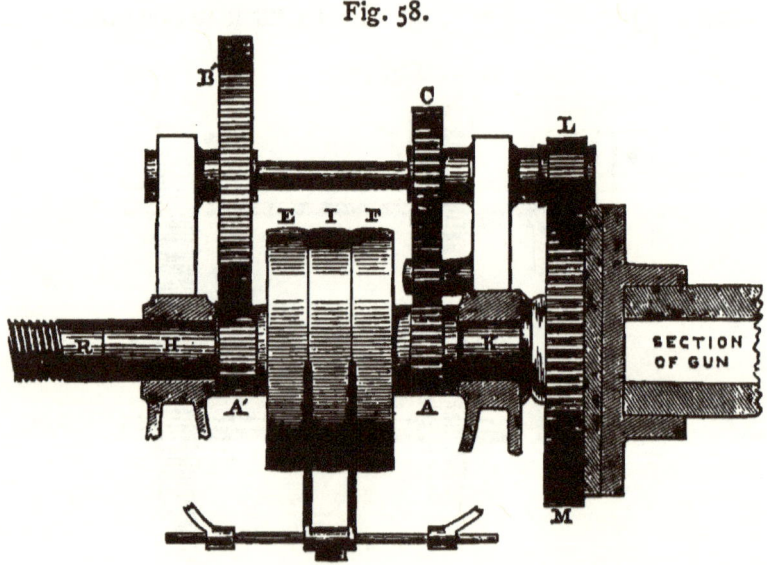

Fig. 58.

When the strap is upon E, the motion travels from A' to B', and so on to L and M, causing the gun and the shaft R K to rotate together slowly in one direction; whereas, upon shifting the strap to F, the motion passes from A to C through a small intermediate wheel, and thence to L and M, whereby the rotation of H K, and of the gun also, is reversed, and a higher speed is introduced.

The object of the machine is to copy upon the interior of

the breech of the gun a screw-thread which is formed upon the shaft, R.

For this purpose the shaft H K is screwed at one end, R, and a slide-rest carrying a cutter advances longitudinally along the gun, with a motion derived directly from a nut which travels along the screw-thread formed upon R. Since the cutter can only remove the metal while passing in one direction, there is a loss of time during the return motion, which it is the object of this combination to reduce as much as possible.

34. We will now examine another class of reversing motions, and we will commence in the most elementary manner.

Conceive a disc E, having a flat edge, to run between two parallel bars, A B and C D, arranged in a rectangular frame; and conceive, further, that the frame can be raised or depressed so as to bring A B and C D alternately into contact with E (Fig. 59).

If the disc rotates always in the direction of the arrows, it will move the frame to the left when brought into contact with C D, and to the right when brought into contact with A B.

We have therefore a reversing motion within the limits of the frame.

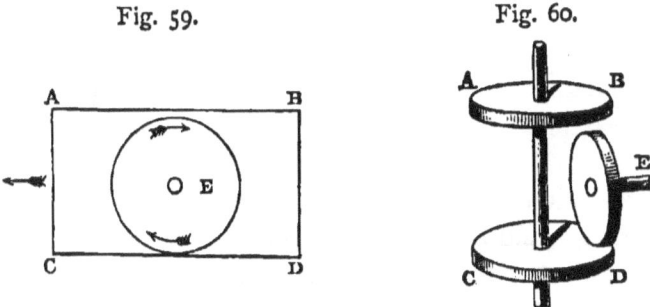

Fig. 59. Fig. 60.

In order to make the motion continuous, it will only be necessary to alter the bars into circular strips or discs, as shown in Fig. 60, and we shall reverse the motion of the vertical axis by bringing the upper or lower discs A B and C D alternately into close contact with the driver E.

In this way we obtain the first idea of a reversing motion, and it only remains for us to improve the general construction and arrangement of the working parts so as to make it practically useful. And we should observe that inasmuch as the rolling action of cones is more perfect than that of circular discs, for the reasons already explained in the introductory chapter, it will be better to substitute cones for the discs, in the manner shown in Fig. 61, and the reversal will occur, just as before, when A B and C D are alternately brought into frictional contact with the driving cone E.

The geometrical condition of rolling will demand that the vertex of the driving cone E shall coincide with that of A B in one position of contact, and with that of C D in the other; hence the vertices of the two cones A B and C D must be separated through a small space equal to that through which the common axis is shifted.

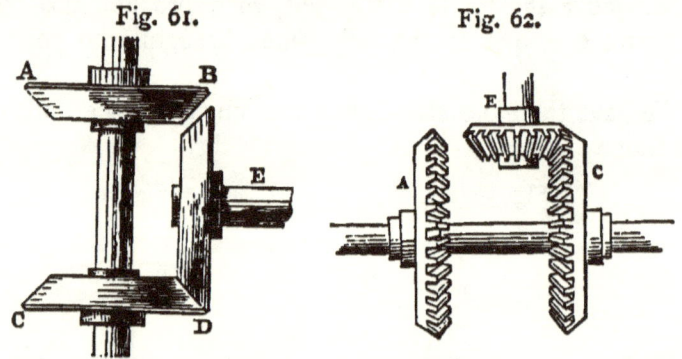

Fig. 61.   Fig. 62.

If we desire to transmit force beyond the limit at which the cones would begin to slip upon each other, we must put teeth upon the rolling surfaces, as in Fig. 62, and we thus obtain a reversing motion which is frequently used in spinning machinery.

Here a bevil wheel, E, is placed between two wheels, A and C, which are keyed to the shaft whose motion is to be reversed, the interval between A and C being enlarged so that E can only be in gear with one of these wheels at the same time; the reversal is then effected by shifting the piece A C longitudinally, so as to allow E to engage with A and C alternately.

35. A reversing motion which depends upon the shifting of wheels in and out of gear is not perfect as a piece of mechanism; we must try, therefore, to convert it into another, so arranged as to give the reversal by passing a driving clutch from one wheel to the other, the wheels concerned in the movement remaining continually in gear and fixed in position.

For this purpose we employ one working pulley F, keyed upon the shaft E D; by its side we place a second pulley I which rides loose upon the shaft, and which carries the driving band when no work is being done.

The wheels A and C ride loose upon the shaft E D, and the intention is to impart the motion of the shaft E D, which is driven by steam power, to the wheels A and C alternately.

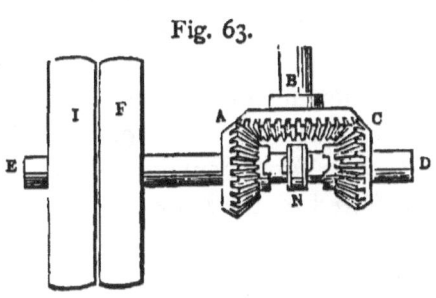

Fig. 63.

We next fit upon the shaft E a sliding clutch N, having projections which serve to lock it to A or C as required, and we place also a projection, or feather, upon the inner part of the clutch which slides in a corresponding groove formed in the shaft, so that N must always turn with E D. It is clear, therefore, that if we allow the clutch N to engage with A we shall communicate to B a rotation in one direction, and that, further, we shall reverse the rotation of B if we connect C with N, for the student will see that in this combination A and C must always rotate in opposite directions, and that the rotation of B as derived from A must be different from that which B would derive from C.

This reversing motion may be commonly seen in steam cranes. The shaft E D is then driven directly by a steam-engine attached to the crane, and the sliding clutch may be locked to either bevil wheel by a friction cone, and is pushed to the right or left by means of a lever which grasps it without preventing its rotation.

There is another application in screwing machines where a rapid reversal is required. In this case the shaft E D is reversed by the action of the bevil wheels, instead of imparting its rotation to each of them in turn. The driving pulley F being attached by a pipe to the wheel A, the reversal is effected by shifting the clutch, and thereby locking the shaft E D to the wheels A and C alternately.

36. We have now to examine the application of this reversing motion in planing machines, and shall describe the combination of three pulleys with three bevil wheels which has been adopted by Sir J. Whitworth.

In pursuing our enquiry into machinery of this character we may remark that the principle of machine copying, whereby a form contained in the apparatus itself is directly transferred to the material to be operated upon, is the distinguishing feature of all planing machines. The application of this principle is perfectly general, and, as a rule, wherever a process of shaping or moulding is well and cheaply performed by the aid of machinery, we find that some skilful and carefully arranged contrivance for transferring a definite form is contained within the machine.

In the earliest form of planing machine, a method of carrying the cutter along parallel bars was adopted, and the present practice is to employ perfectly level and plane surfaces called Vs, which are placed on either side of the machines, and are shaped exactly as their name indicates; their form gives a support to the table, prevents any lateral motion, and allows the oil required for lubrication to remain in a groove at the bottom, from whence it may be worked up by the action of the machine. The table has projecting and similar Vs which rest upon the former, and the object of the mechanism is primarily to cause the table to move in either direction along the grooves, and thus to copy upon a piece of iron supported thereon, and carefully bolted down, an exact plane surface which possesses the truth of the guiding planes.

Whether it may be better to move the table by a rack and pinion or by a screw has been a subject upon which different opinions are held, and at all events the quick return movement, which is given by a combination of spur wheels, as already described in Art. 31, is extremely valuable.

To recur to Sir J. Whitworth's arrangement, we find that he effects the required movement by rotating a screw which runs along the central line of the bed, and which imparts to the table a perfectly smooth traversing motion, equal of course in exactness, if not superior, to that which could be obtained by the best-constructed wheel-work.

There are now three pulleys, E, I, and F, whereof I is an idle pulley, and rides loose upon the shaft; E is keyed to a shaft terminating in the bevil wheel C, and F fits upon a pipe through which the shaft connecting E and C passes, and which terminates in the bevil wheel A.

B is a bevil wheel at the end of the shaft whose direction of rotation is to be reversed.

Fig. 64.

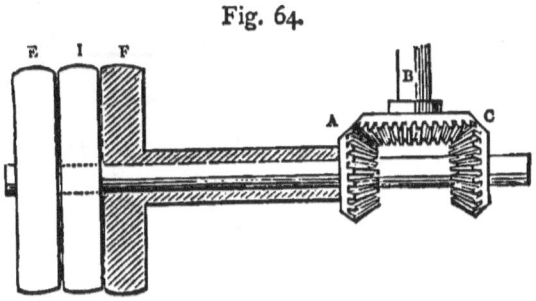

It is clear that the motion of the wheel B is reversed when the driving strap is shifted from E to F.

One objection to this movement consists in the fact that

it does not permit the motion of B to be more rapid in one direction than in the other, and in order to economise the steam-power to the fullest extent, a method of rotating the tool-box was adopted by which means the cut was made while the table traversed in either direction. This reversal answers very well in planing ordinary flat surfaces.

It may, however, be so arranged as to obtain a quick return by making A and C of equal size, and by causing them to gear respectively with two *unequal* wheels upon the axis of B.

37. The contrivance just described is shown in Fig. 65 as applied in a machine for rifling guns, and the method adopted is precisely that so generally employed in planing machines.

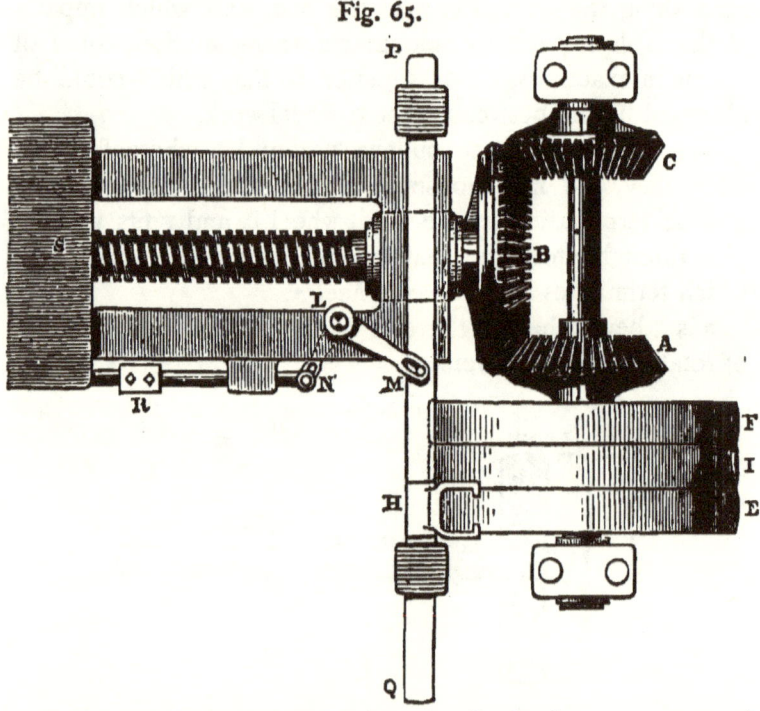

Fig. 65.

The three pulleys and the three bevil wheels are connected together in the manner already indicated, and the bevil wheel B, by its rotation, causes a saddle s carrying the rifling bar to move along the screw in the direction of its length. A bell crank lever, M L N, controls the bar P Q, which carries a fork used to shift the strap, the arms of the lever lying in different horizontal planes; while a moveable piece, R, fixed at any required point of the bar N R, is caught by a projection on the saddle as it passes to the right hand, and thus the bell crank lever is actuated, and the strap is carried along from E to F.

A weight falls over when this is taking place, and gives the motion with sharpness and decision, so as to prevent the strap from resting upon I during its passage. On the return of the saddle to the other end of its path, a similar projection again catches a second piece upon the sliding bar N R, and the strap is thrown back from F to E.

This bell crank lever, as employed for shifting the strap, is worthy of notice; it consists of two arms, L N and L M, lying in different planes and standing out perpendicularly to an axis. It is a contrivance which affords a ready means of transferring a motion from one line, R N, to another, P M Q, which lies in a perpendicular direction at some little distance above it.

38. Where the reciprocation is effected by a contrivance external to the machine itself, two driving bands may be employed: of these one is crossed, and the other is open, as shown in Fig. 66, and it is apparent that the lower discs B and D will turn in opposite directions, in the manner pointed out by the arrows, although they derive their motion from A and C, which, being driven directly by the engine, must rotate in one direction.

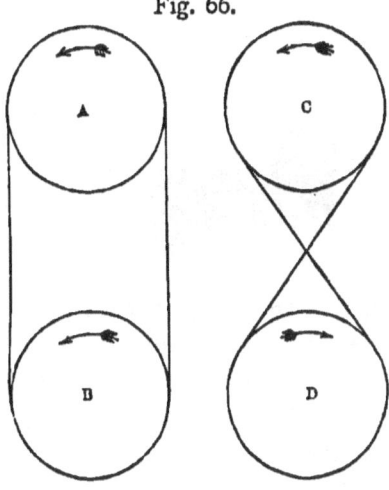

Fig. 66.

The form which the arrangement assumes in practice is shown in the sketch, where one of the driving bands is represented as crossed, and the rotation of the shaft, H K, is derived from each band alternately.

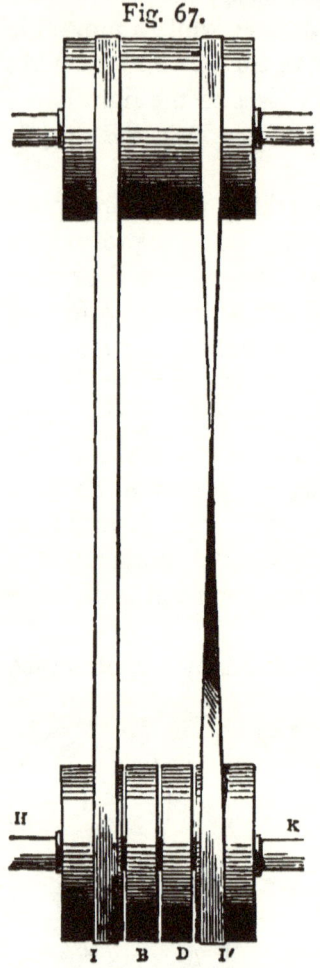

Fig. 67.

There are four pulleys, whereof I and I' are loose upon the shaft, and are twice as broad as B and D, which are the working pulleys.

The bands are shifted by two forks, and remain always at the same distance from each other. In the figure the bands are both upon the idle pulleys, and H K remains at rest. When the bands are shifted to the right, the open strap drives B, and the crossed one remains upon I': when the bands are shifted to the left, the crossed strap drives D, and the open one will be found upon the pulley I. Thus the shifting of the bands will effect the required reversal of the shaft H K.

The motion in either direction is the same, but it may be varied by placing drums of unequal size upon the upper shaft, and again by introducing an inequality between the pair I, B, and the pair I', D. In order to make the explanation clearer, we have described the pulleys B and D as distinct from each other; it is usual, however, to replace them by a single pulley.

39. *Mangle wheels* form a separate class of contrivances for the conversion of circular into reciprocating motion.

A mangle wheel is usually a flat plate or disc furnished

Mangle Motion.

with pins projecting from its face; these pins do not fill up an entire circle upon the wheel, but an interval is left, as shown at F, and E.

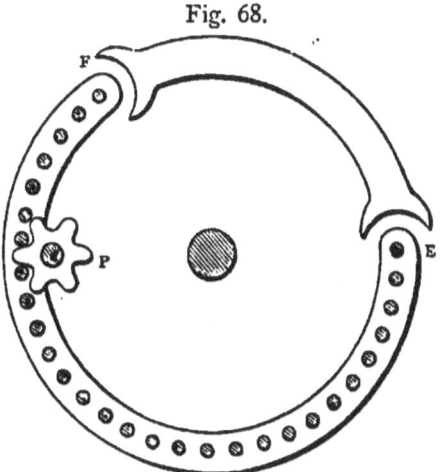

Fig. 68.

A pinion, P, engages with the pins, and is supported in such a manner as to allow of its shifting from the inside to the outside, or conversely, by running round the pins at the openings F and E.

The pinion, P, always turns in the same direction, and the direction of rotation of the mangle wheel is the same as that of P when the pinion is inside the circular arc, and in the opposite direction when the pinion passes to the outside.

By referring to the chapter upon the Teeth of Wheels, we may see that the inner and outer pitch circles coincide in the case of a pin wheel, and therefore that the relative rotation of the mangle wheel to the pinion is precisely the same in both directions.

If the pins be replaced by a curved ring furnished with teeth, the mangle wheel will move more rapidly when the pinion is upon the inside circumference, and by giving certain arbitrary forms to this annulus, the velocities of advance and return may be modified at pleasure. Contrivances such as this are seldom met with at the present time.

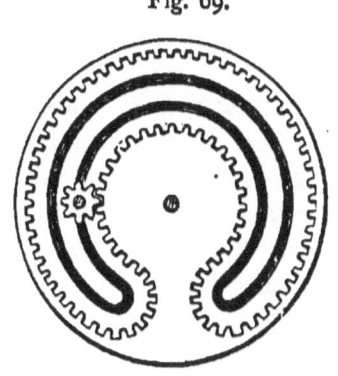

Fig. 69.

40. The mangle wheel may be converted into a *mangle rack* by placing the pins or teeth in a straight line. Here the pinion must be so suspended as to allow of its shifting from the upper to the under side of the rack.

Fig. 70.

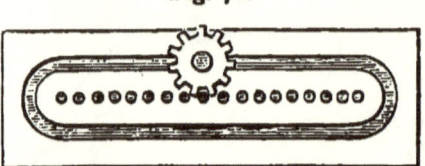

41. Sometimes the pinion is fixed, and the rack shifts laterally; an excellent form of this arrangement was introduced by Mr. Cowper, and serves to give a reciprocating movement to the table in his printing machine.

Fig. 71.

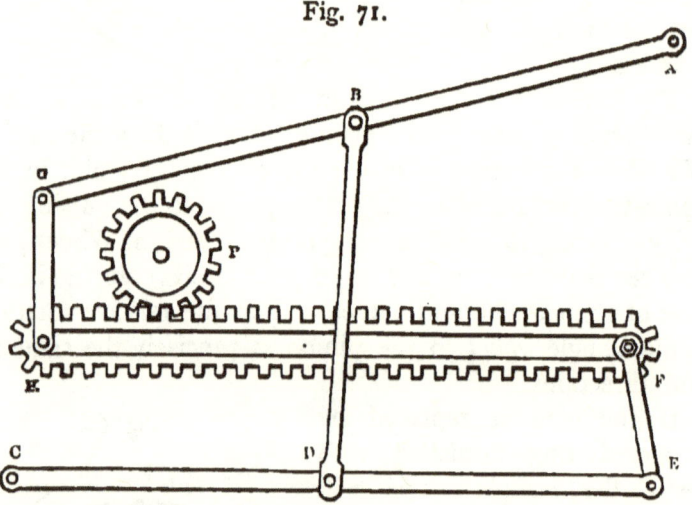

The rack, H F, is attached to the system of bars in the manner exhibited in the diagram. A and C are centres of motion, and are the points where the bars are attached to the table. A G and C E are bisected in B and D, and are joined by the rod B D; the rack H F is attached to the bars A G and C E, by the connecting links, G H and F E, and it must be understood that it is the intention to obtain this so-called mangle motion by the reverse process of fixing the pinion and causing it to drive a continuous rack which runs alternately upon either side of it.

The precise value of the contrivance consists in the arrangement of the bars, which will be understood upon referring to the section upon *Parallel Motion*, and it will be seen that when the pinion has pushed the rack to either end of its path, the bars will so act as to move together, and will shift this rack H F to the opposite side of the pinion, without allowing it to deviate from a direction coincident with that in which the table is moving.

This is the object of the contrivance, and, as we have said, the method by which the result is arrived at will be apparent when the subject of parallel motion has been examined.

Thus the table carrying the parallel bars and the rack oscillates backwards and forwards, while the pinion, which transmits the force, remains fixed in space.

When this machine was applied to the printing of newspapers, the table moved at the rate of 70 inches in a second, and its weight, including the form of type, would be about a ton and a half. When urged to its highest speed the machine would give 5,500 impressions in an hour, which is about the greatest number attainable under a construction of this kind; the true principle in rapid printing being that announced in the year 1790 by Mr. Nicholson, who proposed to place the type upon a cylinder having a continuous circular motion, and upon which another cylinder holding the paper should roll to obtain the impression. But although Mr. Nicholson enunciated the principle more than seventy years ago, and took out a patent for a mode of carrying it out, there is a wide difference between saying that a thing ought to be done, and showing the world how to do it in a practicable manner; hence it was not until late years that Mr. Applegath, and finally Mr. Hoe, were enabled so to arrange their cylinder printing machines upon the principle of continuous circular motion as to satisfy the wants of the 'Times' and the 'Daily Telegraph,' and to print some twelve or fourteen thousand sheets in an hour.

To recur to our shifting rack, it must be remarked that

by reason of the great weight of the table, and the rapidity with which it moves, it would be quite unsafe to leave the rack and pinion in the present unassisted condition; a guide roller therefore determines the position of the pinion relatively to the rack, while the rack itself shifts laterally between guides.

But since, theoretically, the rods would cause H F to move always in a direction parallel to itself; so, practically, they enforce the desired movement in the path of the guides, with as little loss of power as possible.

42. If it be required that the reciprocation shall be intermittent, i.e. that there shall be intervals of rest between each oscillation, we may employ a segmental-wheel and a double rack, as shown in Fig. 72.

The teeth upon the pinion engage alternately with those upon either side of the sliding frame A B, and the motion is of the character required. The intervals of rest are equal, and are separated by *equal* periods of time.

A pin upon the wheel and a guide upon the rack will ensure the due engagement of the teeth.

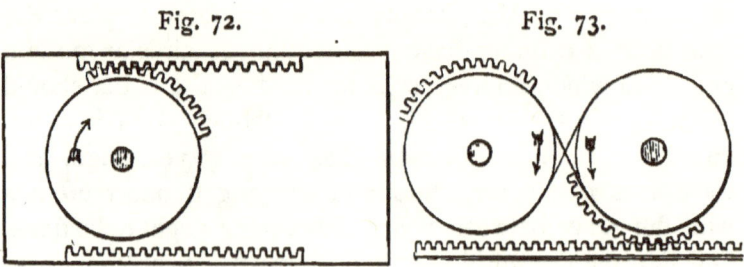

Fig. 72.   Fig. 73.

A mechanical equivalent to the above is found in the use of two segmental wheels and a single rack.

These segments must be equal, but they may be placed in different relative positions upon the discs to which they are attached; and, as a consequence, the intervals of rest may be separated by *unequal* periods of time. (Fig. 73.)

43. These segmental wheels have been employed in the

earlier days of mechanism, and there was a well-known instance in Mr. Cowper's printing machine, where a segment of a wheel engaged with a small sector at each revolution, and so fed on the sheets of paper by the push given while the segments were in action.

Mr. Whitworth has also proposed the subjoined arrangement for the reversal in a machine for cutting screws: we take it as a further example of the use of these wheels, which, however, should always be avoided if possible. There is only one driving pulley, and two segmental wheels are keyed upon the driving shaft. They are close together in the machine, and for the sake of the explanation we have placed one above the other. The object is to effect the reversal of a shaft C: the segmental wheels A and A' have teeth formed round one half of each circumference, and the toothed segments are in situations opposite to each other, as in Fig. 74.

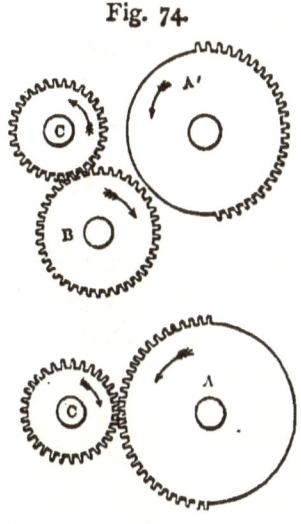

Fig. 74.

When the action of A ceases, that of A' begins, and we have the wheels A and C, or the wheels A', B, and C alternately in action, i. e. we have a reciprocation of C. This is a direct example of the case given in Art. 31.

44. There is yet one most ancient contrivance for changing circular into reciprocating motion, which will repay the trouble of analysing it. In the form exhibited in the next article it was used by the early Greek astronomers to represent mechanically the motion of the Moon.

Let an arm, C P, be centred at C, and convey motion to the grooved arm, B R, by means of a pin, P, which fits into the groove. As C P revolves with a uniform velocity, and in a direction opposite to the hands of a watch, it will cause B R to swing up and down to equal distances upon either side of the line B C, but with this peculiarity, that the upward swing will occupy less time than the downward swing.

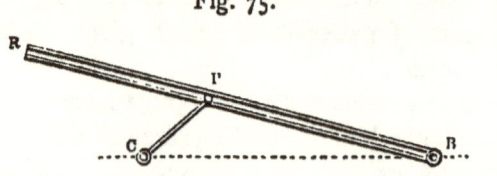

Fig. 75.

The motion of B R will be variable, its velocity changing at every instant, and we must endeavour, in the first instance, to discover an expression for its rate of motion as compared with that of C P. According to well-established rules, we estimate the relative rates of motion of two revolving pieces by comparing the sizes of the small angles described by either piece in a very minute interval of time reckoned from any given instant.

Let now Pp represent the small arc described by P in a very minute interval of time, such as the $\frac{1}{1000}$th part of a second.

Fig. 76.

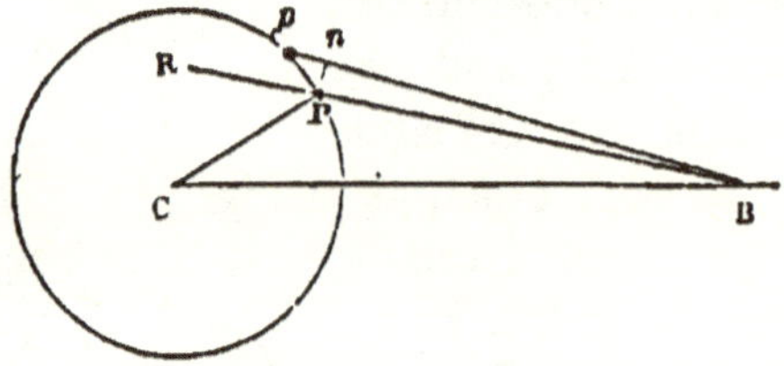

Then
$$\frac{\text{angular vel. of B R}}{\text{angular vel. of C P}} = \frac{\angle\, P B p}{\angle\, P C p}$$

$$= \frac{P n}{B P} \div \frac{P p}{C P}$$

$$= \frac{C P}{B P} \times \frac{P n}{P p}.$$

But $P n = P p \cos p P n$
$ = P p \cos R P C$
$ = P p \cos (C + B)$

$\therefore \dfrac{\angle^{r}\text{ vel. of B R}}{\angle^{r}\text{ vel. of C P}} = \dfrac{C P \cos (C + B)}{P B}.$

Quick Return Motion.

We may test this formula in the usual way; for instance, let C + B = 90°, in which case B R touches the circle, then cos (C + B) = cos 90 = 0;

∴ angular vel. of B R = 0,

or B R stops, as we know it must do.

Next, let C = 0, B = 0, or let P be crossing the line C B, then cos (C + B) = cos . 0 = 1.

$$\therefore \frac{\text{angular vel. of B R}}{\text{angular vel. of C P}} = \frac{\text{C P}}{\text{B P}},$$

or the vel. of B R is as much less than that of C P as B P is greater than C P, a result which is evidently true.

If it be required to find the position of B P when P is at any given point of its path, we have the equation

$$\tan \text{P B C} = \frac{\text{C P} \sin \text{C}}{\text{C B} - \text{C P} \cos \text{C}},$$

whence the angle P B C is known in terms of C.

Note.—The ratio of the angular velocities of B R and C P may also be obtained by analysis.

Let C B P = ϕ, P C B = θ, C P = a, C B = c

Then $\tan \phi = \dfrac{a \sin \theta}{c - a \cos \theta}$

$$\therefore \frac{d\phi}{d\theta} (1 + \tan^2 \phi) = a . \left\{ \frac{(c - a \cos \theta) \cos \theta - a \sin^2 \theta}{(c - a \cos \theta)^2} \right\}$$

$$\therefore \frac{d\phi}{d\theta} . \left\{ \frac{c^2 + a^2 - 2ac \cos \theta}{(c - a \cos \theta)^2} \right\} = a . \left\{ \frac{c \cos \theta - a}{(c - a \cos \theta)^2} \right\}$$

$$\therefore \frac{d\phi}{d\theta} = a . \left\{ \frac{c \cos \theta - a}{\text{P B}^2} \right\}$$

$$= \frac{\text{C P}}{\text{P B}} . \left\{ \frac{\text{C B} \cos \theta - \text{C P}}{\text{P B}} \right\}$$

$$= \frac{\text{C P}}{\text{P B}} . \left\{ \frac{\cos \theta . \sin (\phi + \theta)}{\sin \theta} - \frac{\sin \phi}{\sin \theta} \right\}$$

$$= \frac{\text{C P}}{\text{P B}} . \left\{ \frac{\cos^2 \theta \sin \phi + \cos \phi \cos \theta \sin \theta - \sin \phi}{\sin \theta} \right\}$$

$$= \frac{\text{C P}}{\text{P B}} . \left\{ \frac{\cos \phi \cos \theta . \sin \theta - \sin \phi \sin^2 \theta}{\sin \theta} \right\}$$

$$= \frac{\text{C P}}{\text{P B}} . \cos (\phi + \theta). \quad \text{As in the text.}$$

If we draw B D, B H, tangents to the circle described by P, it will be evident that the times of oscillation of the arm will

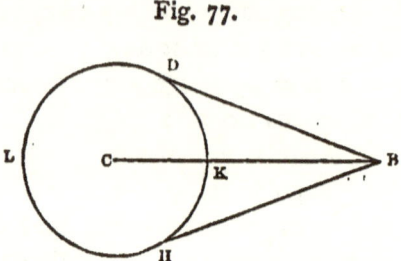

Fig. 77.

be unequal, and will be in the same proportion as the lengths of the arcs D L H, D K H.

45. Hitherto we have supposed C P to be less than C B, and the result has been that B R swings about the point B in unequal times; but we will now arrange that C P shall be greater than C B, in which case B R will sweep completely round with a circular but *variable* motion; we shall, in fact, have solved the problem of making a crank revolve in such a manner that one half of its revolution shall occupy less time than the other half.

Now this is a very important result, and is of great value in machinery, because if the crank be made to perform its two half revolutions in unequal times, it follows that any piece connected with it by a link may be caused to advance slowly and return more rapidly; a movement which, as we have already pointed out, is peculiarly useful in machines for cutting metals.

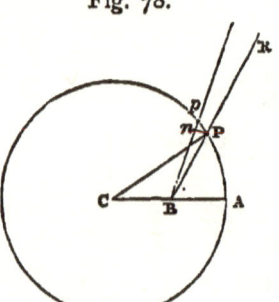

Fig. 78.

As before, let C be the centre of the circle described by P. Then the equation $\tan PBA = \dfrac{CP \sin C}{CP \cos C - CB}$ gives the position of B P when that of the crank is assigned.

Also $\dfrac{\text{angular vel. of B R}}{\text{angular vel. of C P}} = \dfrac{Pn}{PB} \div \dfrac{Pp}{CP} = \dfrac{Pp \cos CPB}{PB} \times \dfrac{CP}{Pp}$

$= \dfrac{CP}{PB} \cdot \cos CPB.$

Cor. 1. If C B be small, the angle C P B will be small also and we shall have $\cos CPB = 1$ nearly; in which case the angular vel. of B R varies as $\dfrac{1}{BP}$, while that of C P remains constant.

Cor. 2. When C B P is a right angle, we have

$\cos CPB = \dfrac{PB}{CP}$, or $\dfrac{CP \cos CPB}{PB} = 1$,

that is, the angular vel. of B R = the angular vel. of C P.

This happens twice during a revolution, and gives the line of division of the inequalities of the motion of B R. Hence, if we draw D B H perpendicular to C B and cutting the circle described by P in the points D and H, the times of each half revolution of B R will be in the proportion of the arcs D K H and D L H.

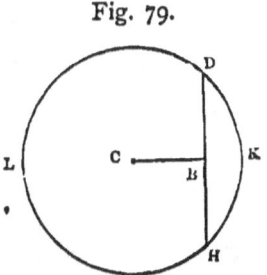

Fig. 79.

46. If B R be made to carry a link, R Q, as in the case of the crank and connecting rod, the linear motion of Q will be the same in amount as if B R revolved uniformly, but the periods of each reciprocation will in general be different.

The difference in the times of oscillation will depend upon the direction of the line in which Q moves.

The best position for that line is in a direction $\perp^r$ to C B; we have shown that the times of oscillation are always as the arcs D K H and D L H, and it is also evident that the inequality between these arcs is greatest when D H is $\perp^r$ to C B, and diminishes to zero when D H passes through C B.

We have now an arrangement very suitable for effecting a quick return of the cutter in a shaping machine.

Let one end of a connecting rod be made to oscillate in a

line $\perp^r$ C B, or nearly so, and let the crank B R be driven by an arm, C P, which revolves uniformly in the direction of the arrow, we at once perceive that Q will advance slowly and return quickly, the periods of advance and return being as the arcs D L H and D K H.

Fig. 80.

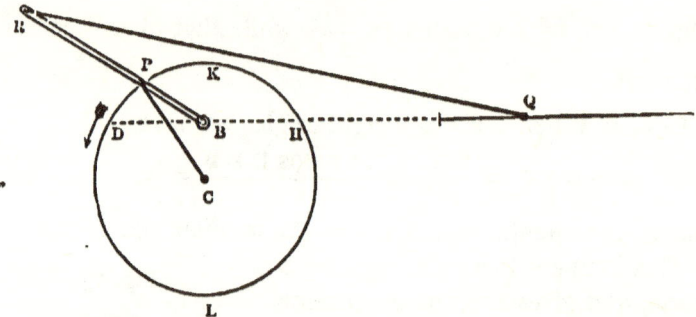

47. Such a direct construction is not very convenient for the transmission of force, and it has been so modified by Sir J. Whitworth in his *Shaping Machine*, that the principle remains unchanged, while the details of the moving parts have undergone some transformation.

This machine is analogous to a planing machine, but there is no moveable table; the piece of metal to be shaped is fixed and the cutter travels over it. The object of the contrivance is to economise time, and to bring the cutter rapidly back again after it has done its work.

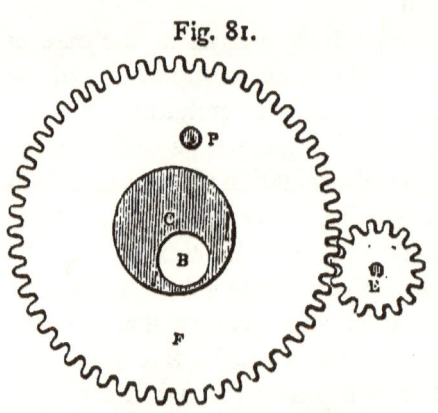

Fig. 81.

The arm C P is here obtained *indirectly* by fixing a pin, P, upon the face of a plate, F, which rides loose upon a shaft, C, and is driven by a pinion, E.

As the wheel F revolves upon the shaft represented by the

shaded circle, the pin moves round with it, and remains at a constant distance from its centre.

A hole, B, is bored in the shaft, C, and serves as a centre of motion for a crank piece, D R, shown in Fig. 82. The connecting rod, R Q, is attached to one side of this crank piece, and the pin, P, works in a groove upon the other side. Thus the rotation of the crank causes the end Q to oscillate backwards and forwards, and to return more rapidly than it advances.

The length of the stroke made by Q must be regulated by the character of the work done, and is made greater or less by shifting R farther from or nearer to B; this adjustment

Fig. 82.

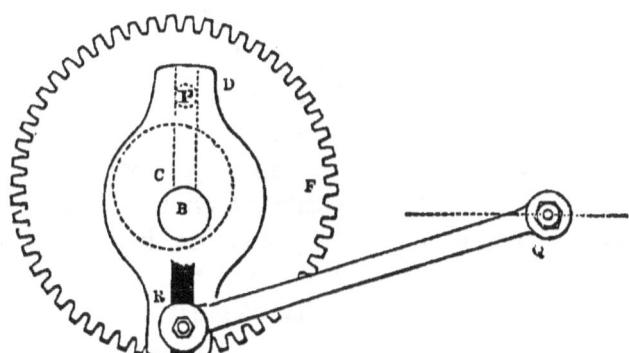

does not affect the inequality in the relation between the periods of advance and return which the machine is intended to produce.

48. As a further illustration of this slit-bar motion, we give a sketch of a curvilinear shaping machine used at the Crewe Locomotive Works.

There have been instances of unequal wear of the tyres in the leading wheels of locomotive engines, which have been traced to the circumstance of the wheel itself being a little out of balance; that is to say, the centre of gravity of the wheel did not exactly coincide with its centre of figure.

In one case a wheel was found upon trial to be 9 lbs. out of balance.

Now we learn in mechanics that a weight of W lbs. describing a circle of radius (r) with a velocity of (v) feet per second, will, during its whole motion, exert continually a pull upon the centre, in the direction of a line joining the body and the centre, which will be measured in pounds by the expression

$$\frac{Wv^2}{32\cdot 2 \times r}.$$

Suppose a wheel 3 ft. 6 in. diameter to run at a velocity of 50 miles an hour, in this case v will be equal to $\frac{220}{3}$, and r will be $\frac{7}{4}$,* whence the pull of only one pound weight at a distance of $3\frac{1}{2}$ ft. from the centre will amount to rather more than 95 lbs., and a weight of 9 lbs. would produce a pressure upon the bearing of rather more than $7\frac{1}{2}$ cwts.; and then, in the time of a half revolution, viz. about $\frac{1}{13}$th part of a second, the same pressure in the opposite direction.

It is, of course, only at high speeds that the defects due to want of balance become serious, and this numerical result shows very plainly the necessity of great care in the construction of wheels which are required to run at a high velocity.

The machine intended to shape the curved inner face of the rim of locomotive wheels has the quick return movement which we have just discussed.

---

* Here $\quad v = \dfrac{50 \times 5280}{60 \times 60} = \dfrac{5 \times 44}{3} = \dfrac{220}{3}$

and $\quad \dfrac{Wv^2}{32\cdot 2 \times r} = \dfrac{1 \times 220 \times 220}{9 \times 32\cdot 2 \times \frac{7}{4}}$

$= \dfrac{8 \times 12100}{7 \times 9 \times 16\cdot 1}$

$= \dfrac{96800}{1014\cdot 3} = 95\cdot 4.$

The point B is the centre of motion of the lever bar, and coincides with the centre of the circular portion forming the

Fig. 83.

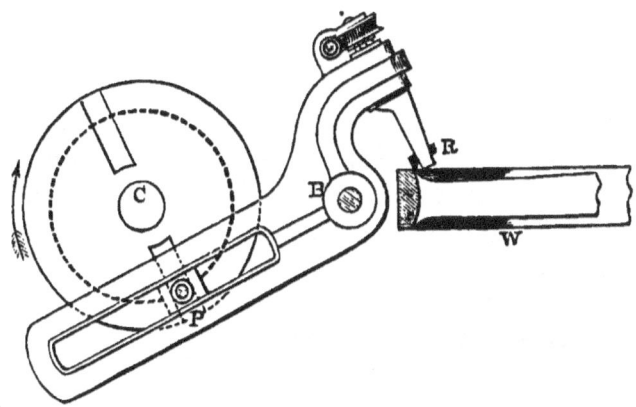

inner surface of the rim of the wheel w. The tail end of the lever has a long slot in which the crank pin P works; this pin is attached to the driving disc centred at C, and the length of the stroke can be adjusted by shifting P in the direction of the radius C P.

49. From the invention of the Art of Printing in the year 1450 till the year 1798 no material improvement was made in the Printing Press. The earliest representation of a press occurs as a device in books printed by Ascensius; there is scarcely any difference between it and a modern press, and it is truly a matter of astonishment that so long a period as nearly 350 years should have rolled on without some improvement being made in so important a machine.

The wooden press consists of two upright pieces of timber joined by transverse pieces at the top and near the bottom; a screw furnished with a lever works into the top piece, and by its descent forces down a block of mahogany, called the 'platten,' and thus presses the sheet of paper upon the type, which is laid upon a smooth slab of stone embedded in a box underneath. In the year 1798 Lord Stanhope constructed the press of iron instead of wood, and at once transferred the machine from the hands of the carpenter to those of the engineer; he further added a beautiful combination of levers for giving motion to the screw, causing thereby the platten to descend with decreasing rapidity, and consequently increasing force, until it reached the type, when a very great power was obtained.

The Stanhope levers consist of a combination of two arms or cranks, C P, B Q (Fig. 84), connected by a link, P Q, in such a manner that the connecting link shall come into a position perpendicular to one of the arms at the instant that it is passing over the centre of motion of the other arm.

In order that this may happen, it is evident that the various pieces must satisfy the relation

$$P Q - C P = \sqrt{C B^2 - B Q^2}.$$

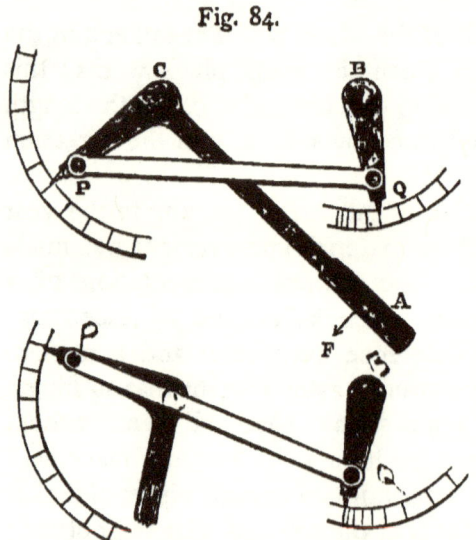

Fig. 84.

For the convenience of the workman who is employed upon the press, a handle, C A, is attached to the crank C P, and moves as part of it, but the introduction of this handle does not affect the principle of the movement, which, regarded as a question in mechanics, depends simply on the combination of C P, B Q, and P Q.

If, now, a force, F, be applied at the end of the handle, A C, so as to turn the crank, C P, uniformly in the direction indicated, the arm, B Q, will, under the conditions already stated, move with a continually decreasing velocity until it comes to rest, and then any further motion of C P will cause B Q to return.

The lower diagram shows the levers in this extreme position, and the graduated scales at P and Q indicate the relative angular movements of C P and B Q.

Now the motion, interpreted with relation to the transmission of force, implies that the resistance at Q necessary to balance the moving power which turns the crank is increasing rapidly as the rotation of B Q decreases, and that there is no limit theoretically to the pressure which will be felt as a pull at Q by reason of the force F. In practice this extreme pressure is exerted through so very small a space that the theoretical advantages are scarcely realised, but the arrangement is exceedingly useful as applied in the Printing Press.

The lever B Q is there employed to turn the screw which acts upon the platten; the workman gives a pull to the handle, A C, and by doing so causes the platten to descend with a motion which is at first considerable, and afterwards rapidly dies away. At the same time, the limited amount of power which is being exerted comes out with greatly magnified effect in impressing the paper upon the type.

50. In order that this contrivance may be thoroughly understood, take the annexed sketch to represent it, and draw C N $\perp^r$ to P Q.

Fig. 85.

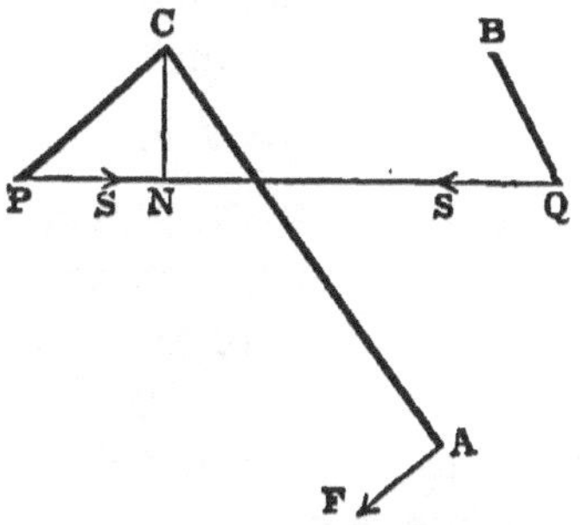

A force F acting at A in a direction $\perp^r$ to C A would be balanced by a force S acting in P Q, such that

$$S \times C N = F \times C A,$$
$$\text{or} \quad S = \frac{F \times C A}{C N}.$$

This force S, necessary to balance F, would be supplied by the resistance to motion in the arm B Q, and would, in fact, be the pull felt at Q. Now as the arms turn, the link P Q gets nearer and nearer to C, and C N becomes less and less until it has no appreciable magnitude, and the consequence is that s increases enormously in the last instant of the motion.

Ex. Let F = 20 lbs., C A = 20 inches, C N = $\frac{1}{10}$th of an inch, we have

$$s = \frac{20 \times 20}{\frac{1}{10}} \text{ lbs.} = 4000 \text{ lbs.}$$

In connection with this subject we may here examine the *knuckle joint* used in the so-called Columbian Printing presses, and also in machinery for punching or shearing iron.

Here two arms C P, P Q, generally equal in length, but which may be unequal, are jointed together as in the sketch, the point C is *fixed*, and the end Q exerts a pressure which may be carried on by a piece moving in the direction C Q.

Fig. 86.

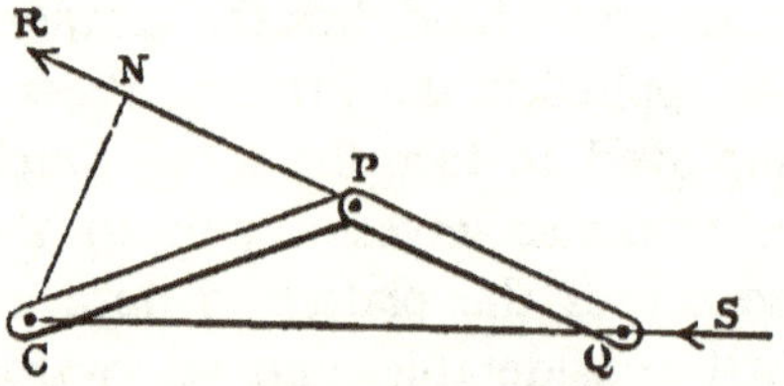

The force F, which produces the result, is supposed to act upon the joint at P, and the reaction s which is felt at Q, will be transmitted also to P in the direction P Q.

Let this reaction, resolved in the line Q P, be equal to R, draw C N $\perp^r$ to P R.

Then, by the principle of the lever, we have

R × C N = F × (perpendicular from C upon F).

But as C P Q straightens, C N diminishes and ultimately becomes equal to zero while the product F × (perpendicular from C upon F) remains constant; it follows therefore that the product of R × C N remains constant while C N diminishes to nothing, but this can only happen if R increases in the same proportion that C N diminishes, that is, without limit. Hence the power of the combination.

It is upon this principle that the heavy chain of a suspension bridge cannot be stretched into a straight line; it would break long before it straightened.

In the same way, if the joint be doubled back, the point Q being *fixed*, and the force F being supposed to act in a line ⊥ʳ to C Q, we shall have the pull upon C, which we may call R, felt as a pressure in the line C P.

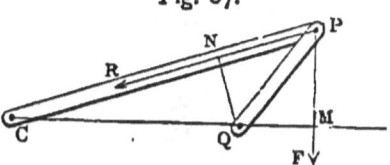

Fig. 87.

If now Q N be drawn ⊥ʳ to the direction of R, as felt at P, the principle of the lever gives us the equation

R × Q N = F × Q M,

and R becomes infinite when Q N vanishes, or when C P is passing over Q.

This is the case of the Stanhope levers.

51. The knuckle joint is often found in combination with the ordinary crank and connecting rod, as shown in the diagram, where the point P describes a circle round C, and causes the joint D Q E to straighten each time that P reaches its lowest position.

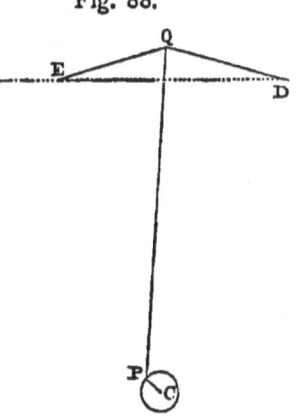

Fig. 88.

In this case C should be in the vertical line passing through Q when the joint is straightened, and P Q should be just long enough to reach the lowest point of the circle at the same instant. It will follow that the joint can, under these circumstances, straighten itself only once during a complete revolution of C P, and the contrivance may then be applied so as to obtain a decreasing motion of the point E, and thereby to transmit a pressure which greatly and rapidly increases.

An example occurs in printing machinery, where a knuckle joint, actuated by a crank exactly as described above, is employed to depress the platten upon the impression table, and so to effect, in a large machine worked by steam power, the same thing which is done on a smaller scale by the pull of the lever of a hand press.

In the case last treated, the point Q never passed below the line D R, and thus the joint only straightened itself once during a revolution of C P; it is possible, however, to cause this straightening to occur twice in each revolut' ' the crank, and to effect this, it is only necessary   .t the point C a little nearer to D Q, so that the joint r.ay straighten when P is upon either side of the lowest or highest point of its circular path.

The sketch shows the knuckle joint as applied to a movement of this character in a Power Loom. (Fig. 89.)

Fig. 89.

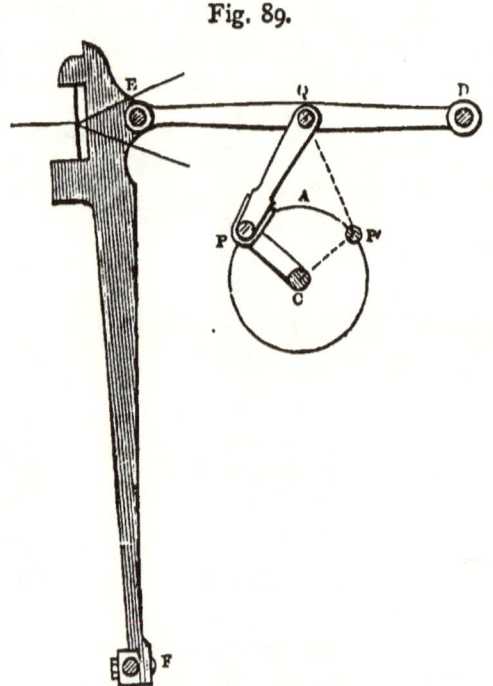

In this case the joint will straighten when P arrives at certain points on either side of the vertical line C A, as shown by the positions of C P, C P', and thus we shall find that the point Q falls below D E, as well as rises above it, and that there will be two positions of P upon either side of A in which D Q E becomes a straight line.

In weaving, the thread of the weft requires to be beaten up into its place after each throw of the shuttle; and in some ~~es,~~ as in carpet weaving, two beats are wanted instead ~~ne.~~

The arrangement which we are now discussing has been used to actuate the moveable swinging frame, or batten, which beats up the weft, and the result is that two blows are given in rapid succession.

In the figure referred to, E F is the batten, moveable about F as a centre, and it is clear that when the crank takes the positions C P, C P', the joint D Q E will straighten, and, as a consequence, the batten will be pushed as far as it can go to the left hand, or a beat-up of the weft will take place.

We thus solve the problem of causing a reciprocating piece to make *two* oscillations for each complete revolution of an arm with which it is connected.

By recurring to the earlier part of the chapter, we shall understand that a cam-plate moveable about C, and shaped as in Fig. 90, may be employed to drive the batten, and may replace the above combination, being, in point of fact, a mechanical equivalent for it.

Fig. 90.

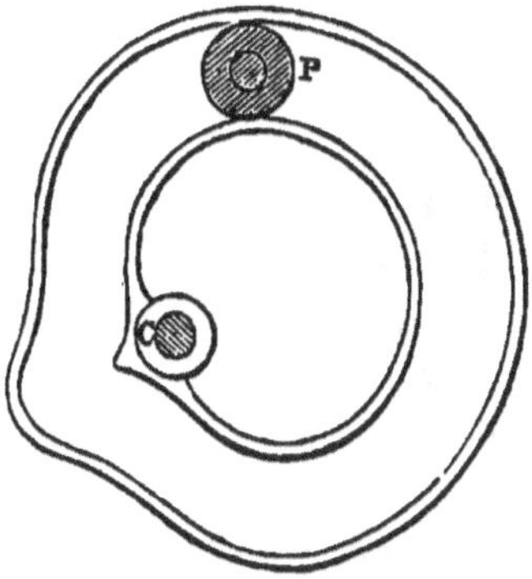

The roller P is then connected with levers attached to the batten, and the beat-up occurs when P passes through the hollows upon each side of the projecting portion of its path upon the plate.

This principle of obtaining *two* vibrations of a bar for each revolution of the driving-crank may be extended still further, and we will alter the construction so as to obtain *four* vibrations instead of two.

The arrangement of the knuckle joint D Q E and the crank C P remains as before, but the arm Q E is now connected with a second knuckle joint F B L, and the piece A K,

Fig. 91.

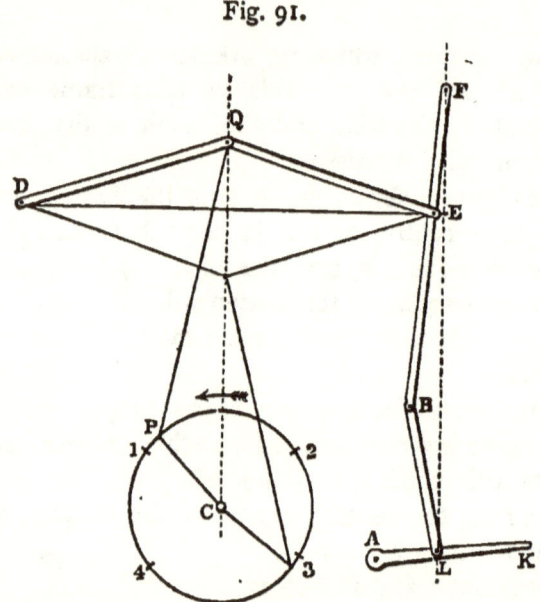

which is to receive *four vibrations*, is centred at A, and is attached at the point L to the link B L.

It is clear that the joint F B L will straighten itself four times in each revolution, viz. when the crank C P is in the positions marked 1, 2, 3, 4, upon the circle, and thus A K will make four complete vibrations for each revolution of the crank.

In other words there are two positions of the joint D Q E in which F B L straightens, and each of these positions of D Q E is obtained by two distinct positions of C P.

52. It is frequently convenient to make the rotation of one crank produce a movement of oscillation in another.

Knuckle Joint.

There is no difficulty in arranging that a small oscillation of C P shall cause B Q also to oscillate; but if it be required that C P shall make complete revolutions, some consideration will be necessary.

Upon referring to Fig. 92, it will be seen that the points E and H, where the oscillation of Q terminates, are given by the condition that C P and P Q fall into the same straight line.

$$\therefore \text{C E} = \text{P Q} - \text{C P}$$
$$\text{C H} = \text{P Q} + \text{C P}.$$

If C P is to make complete revolutions, instead of merely reciprocating, it will be necessary that C P and P Q shall come into a straight line before B Q and P Q have the power to do so.

Fig. 92.

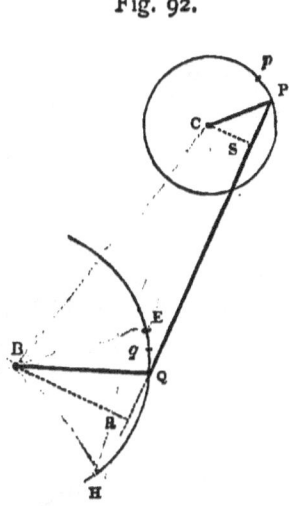

Hence the required conditions are that

$$\text{C B} + \text{B Q} > \text{C H}$$
$$\text{or} > \text{P Q} + \text{C P}$$
$$\text{and C B} - \text{B Q} < \text{C E}$$
$$\text{or} < \text{P Q} - \text{C P}.$$

It will be readily seen, upon testing this statement, that if C B be taken equal to P Q, the crank C P will revolve and B Q will oscillate, so long as C P is sensibly less than B Q.

The angle of swing of B Q increases also as C P becomes more nearly equal to B Q, and tends to reach two right angles as a limit.

53. Finally, we have to compare the angular velocities of C P and B Q; from this comparison we could, by the principle of work, infer the relation between any forces which might be transmitted from the one crank to the other in respect of magnitude.

Draw C S, B R perpendiculars upon the direction of P Q,

and let P and Q shift to p and q during the smallest conceivable interval at the beginning of the motion.

Then the resolved part of the motion of Q in the direction Q P is equal to $Qq \cos qQP$,

$$= Qq \sin BQR$$
$$= Qq \times \frac{BR}{BQ}$$
$$= BR \times \frac{Qq}{BQ}$$
$$= BR \times \text{angle } QBq.$$

So also the resolved part of the motion of P in the direction $QP = CS \times \text{angle } PCp$.

But in the first instant of the motion these resolved parts are equal to each other, because P Q remains for a brief space parallel to its first position.

$$\therefore BR \times \text{angle } QBq = CS \times \text{angle } PCp,$$
$$\therefore \frac{\text{angle } QBq}{\text{angle } PCp} = \frac{CS}{BR}.$$

But although the angular velocities of the arms B Q and C P change continuously, yet they will be at any instant in the same proportion as the angles described by these arms in a very minute interval of time, the relative motions of the arms not being supposed to change during that interval.

Hence
$$\frac{\text{angular vel. of } BQ}{\text{angular vel. of } CP} = \frac{\text{angle } QBq}{\text{angle } PCp}$$
$$= \frac{CS}{BR}.$$

With this proposition we conclude the present chapter, and we shall recur to a comparison of the angular velocities of two arms connected together by a link when we discuss the forms to be given to the teeth of wheels.

CHAPTER II.

ON THE CONVERSION OF RECIPROCATING INTO CIRCULAR MOTION.

54. It has been shown that the motion of a point in a circle results from the combination of two reciprocating movements in lines at right angles to one another, and that circular may be converted into reciprocating motion by the suppression of one of these movements.

The reconversion of reciprocating into circular motion is not a problem of the same kind, as we now require the creation of a movement, instead of its suppression: such a creation is impossible in a strict mathematical sense, but is practically attainable by mechanical construction.

Fig. 93.

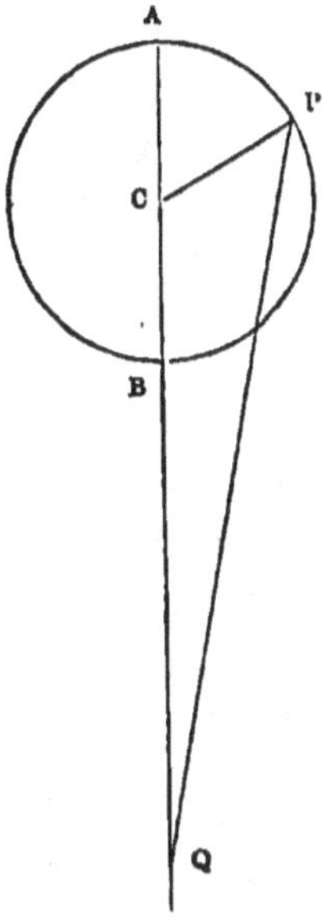

We recur now to the contrivance of the crank and connecting rod as one of the most obvious methods of solving the problem; it is clear that the travel of Q in a line pointing to C will cause the rotation of C P, and will compel P to move in a circular arc.

But unless C P possesses inertia, or is attached to some heavy body as a flywheel, which, when once set in motion, cannot suddenly come to rest, there will be two points where the power exerted at Q will fail to continue the motion, and these points are evidently at A and B, where C P Q straightens into a right line.

It is usual to call A and B the *dead points* in the motion, and P must be made to pass through them without deriving any aid from Q.

In the existence of these *dead points* we note the failure of the contrivance as a piece of pure mechanism, and theory would tell us beforehand that it must fail, because you cannot create motion by mechanism any more than you can create force: you may modify or interchange without limit the movements which exist among the parts of a system, but there your power ends, and accordingly the continuous rotation of C P can only be provided for by storing up in the arm itself, or in some body, such as a fly-wheel, connected with it, the energy which is necessary to overcome any external resistance during that portion of the movement where the driver ceases to act.

So, therefore, in applying the crank and connecting rod to beam-engines, the piston rod is attached by Watts' parallel motion to one end of a heavy iron beam, and the rotation of the fly-wheel is derived by the aid of a connecting rod or spear uniting the other end of the beam with a crank which turns the fly-wheel.

The application to direct acting engines, of which the locomotive engine may be taken as the type, has been already noticed, and the student will now understand that the mechanical working would be incomplete unless the crank were attached to a fly-wheel or other heavy revolving body balanced upon its centre, which would carry P through those portions of its path near to the dead points, and would also act as a reservoir, into which the work done by the steam might be poured, as it were, unequally, and from which it might be drawn off uniformly, so as to cause the engine to move smoothly and evenly.

In marine engines, where the fly-wheel is not admissible, and where the engine must admit of being readily started in any position, two separate and independent pistons give motion to the crank shaft. In this case the two cranks are

Ratchet Wheels.

placed at right angles to each other, so that when one crank is in a bad position the other is in a good one.

The same plan holds in the construction of locomotive engines.

55. In the instances considered, the circular motion derived from the reciprocating piece is *continuous*; it now remains for us to examine a class of contrivances for producing the like result where the circular motion is *intermittent*.

The circular motion being that of a wheel turning upon its axis, it may be arranged that one-half of the reciprocating movement shall be suppressed, and that the other half shall always push the wheel in the same direction; this is the principle of the *ratchet wheel*.

Or it may be arranged that the reciprocating piece shall be of the form shown in those escapements which produce a recoil, and the pallets will then act upon opposite sides of the wheel, and will push it always in one uniform direction. Here we find ourselves again in the region of pure mechanism. These two classes of contrivances constitute the only methods by which the problem can be solved without external assistance.

56. A wheel provided with pins or teeth of a suitable form, and which receives an intermittent circular motion from some vibrating piece, is called a *ratchet wheel*.

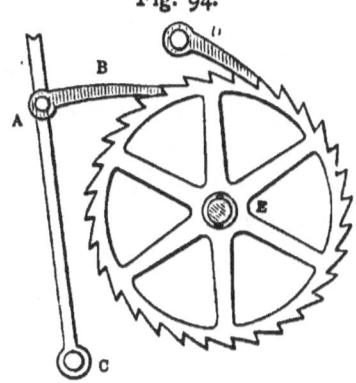

Fig. 94.

In the drawing E represents the ratchet wheel furnished with teeth shaped like those of a saw, and A B, the driver, is a click or paul, jointed at one end A, to a moveable arm A C, which has a vibrating motion upon C as a centre.

As A C moves to the right hand, the click, B, pushes the

wheel before it through a certain space ; upon the return of A C, the click, B, slides over the points of the teeth, and is ready again to push the wheel through the same space as before, being in all cases pressed against the teeth by its weight or by a spring.

A detent, D, prevents the wheel from receding, while B is moving over the teeth, for it is, of course, a condition in this movement that the ratchet wheel itself shall either tend always to fly back, or shall remain held in its place by the friction of the pieces with which it is connected.

In this way the reciprocating movement of A B is rendered inoperative in one direction, and the circular motion results from the suppression of one half of the reciprocating movement of the arm.

The wheel, E, and the vibrating arm, A C, are often centred upon the same axis.

The usual form of the teeth is that given in Fig. 94, and the result is that the wheel can only be driven in one

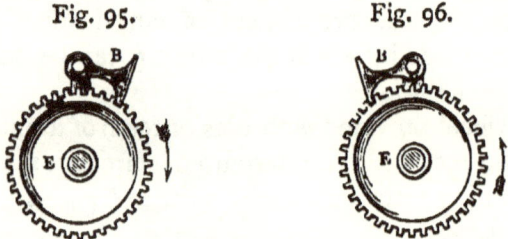

Fig. 95. Fig. 96.

direction ; but in machinery for cutting metals it is frequently desirable to drive the wheel indifferently in either direction ; in that case the annexed construction is adopted. The ratchet wheel has radial teeth, and the click, B, can take the two positions shown in the diagrams, and can drive the wheel in opposite directions. (Figs. 95, 96.)

In this case the click has a triangular piece upon its axis, any side of which can be held quite firmly by a flat stop attached to a spring ; there are, therefore, three positions of rest for the click, whereof two are shown in the figure, and the third would be found when the click was thrown up in the direction of the arm produced. We have by this simple contrivance a ready means, not only of driving the wheel in both directions, but also of throwing the click out of gear when required.

57. As regards the action between the teeth and the detent, we observe that the wheel must tend to hold the detent down by the pressure which it exerts, and that it will do so as long as the line of pressure on the surface, pr, falls below the centre D.

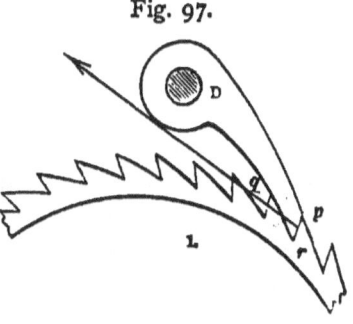

Fig. 97.

If the angle qrp were opened out much more, the perpendicular upon pr might rise above D, and the detent would then fail to hold the wheel.

Further, the click has to return by slipping over the points of the teeth; the condition for this result is that the perpendicular to the surface qr shall fall between D and the centre of the wheel.

Where very little force is required to hold the wheel, and the exact position is of consequence, as in counting machinery, the teeth may be pins, and the detent may be a roller pressed against them by a spring.

58. Everyone must have seen the application of the ratchet wheel to capstans and windlasses, where it is introduced in order to prevent the recoil of the barrel; the same purpose for which it is applied in clocks and watches.

It was a very early improvement to provide two pauls of different lengths, termed by the sailors *paul and half paul*, and thereby to hold up the barrel at shorter intervals during the winding on of the rope; in fact, a ratchet wheel of eight teeth thus became practically equivalent to one of sixteen teeth, and the men were better protected from any injury which might be caused by the recoil of their handspikes.

The principle of this contrivance is very intelligible, and is shown in the diagram, where the two pauls D P, E Q, differ in length by half the space of a tooth.

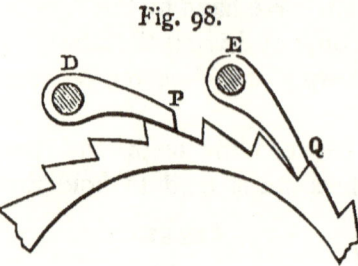

Fig. 98.

As the wheel advances by intervals of half a tooth, each paul falls alternately, and the same effect is produced as if the number of teeth were doubled, and there was one paul.

In the same way three pauls might be used, each differing in length by one-third of the space of a tooth, and so the subdivision might be extended.

We come now, by an easy step, to a contrivance which may often prove valuable.

In practice you may be required to move a ratchet wheel through certain exact spaces, differing by small intervals; where such is the case it is better not to attempt a minute subdivision of the teeth, as they become liable to break and wear away and the action is uncertain, but recourse should be had to a method of placing three or four clicks upon the driving arm.

In such a case the pauls or clicks will be increased in number, and will act as drivers instead of detents, being arranged in order of magnitude as regards length. They may be placed upon separate driving arms, but there is no advantage in doing so, and it is usual to place them all upon a pin at the end of the driving bar.

In the annexed figure two clicks D P, D Q, are shown hung upon a pin D, which is supposed to be at the end of the arm which drives the wheel; the clicks differ in length by half the space of a tooth, and they will manifestly engage the wheel alternately, and

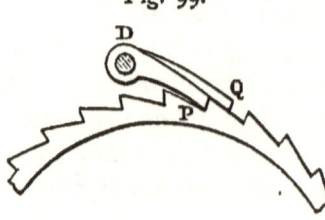

Fig. 99.

will move it as if there were twice as many teeth driven by one click. And so for *three* or any greater number of clicks.

On referring back to Fig. 24, the student will find an example of the use of a ratchet wheel. A link, H K, connects the reciprocating frame, F G, with an arm, L M, carrying a click at Q; thus the oscillations of the frame are received by the arm, and the wheel is advanced a certain number of teeth upon each motion to the right. The number of teeth taken up can be regulated by adjusting the distance of K from L by means of a screw; the nearer K is brought to L, the greater will be the advance of the ratchet wheel at each stroke.

The object of the arrangement is to feed on the rod of lead from which the material for each bullet is cut, and by placing *three* clicks at Q instead of one, according to the method just examined, the amount of advance for bullets of different sizes may be regulated with considerable nicety.

As, in some form or other, the principle of this construction is of great practical value, we will examine it more in detail. It will be found that one chief use of the ratchet wheel occurs in providing the feed motions in machinery for cutting or shaping metals, and the general plan adopted is to draw off, as it were, some definite amount of motion at proper intervals during the operation, and so to impart an unchangeable movement of vibration through a definite angle to a bar which comes as the first piece on the way to the ratchet wheel. We start by giving to such a bar some fixed and unvarying amount of vibration, and our object will be to draw off from this motion just so much or so little as we may require for the purposes of the work, and so to advance the ratchet by 1, 2, 3, or any convenient number of teeth.

If we can now arrange to do this, we have obtained the first part of a feed motion, viz. the power of advancing the ratchet at each stroke by an integral number of teeth; but we may want to go further, and advance the wheel with greater nicety, such as through 2⅓ of a tooth each time, and the student will now understand that for the integral part of the advance some definite contrivance, such as those we are about to discuss, will be wanted, and that for the fractional part of the advance this system of multiple clicks will be perfectly sufficient.

59. In considering this problem of advancing the ratchet through any required number of teeth, a very obvious principle will no doubt suggest itself which may be understood at once upon inspecting the diagram, Fig. 100.

Suppose the arm A C to represent a vibrating bar which swings with a definite amount of angular motion through the space P C Q. In a planing machine this bar would be pushed over each time that the table came to the end of its travel. If A C be connected to another point S by the link R S, we can impart to the point S a reciprocating motion in the line C S, which may be represented by the space F E under the conditions shown in the sketch. It is now clear that if R be moved towards C, this travel, F E, will diminish to nothing, while A C continues all the time to swing through the same angle P C Q, whereas F E will be increased in a like degree when R is moved away from the centre C in the direction R r.

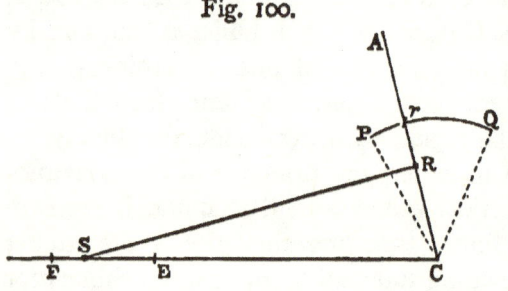

Fig. 100.

Thus the rate of advance of the ratchet wheel may be regulated.

60. Another method, extremely ingenious in principle, has been applied in Sir J. Whitworth's planing machines, and is worth a careful examination. In Fig. 101, let C'A represent a vibrating bar centred at C, upon which point there is also centred a circle carrying two pins, P and Q.

Feed Motions.

We will suppose that the circle vibrates, independently of the arm, through an angle exactly represented by P C Q, and that the object of the contrivance is to impart to the bar, C A, the whole or any portion of this vibration.

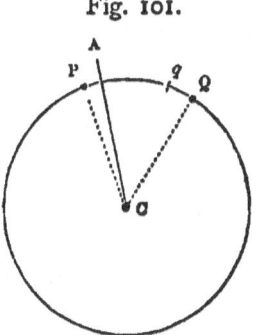

Fig. 101.

It is easy enough to impart the whole vibration; we have only to fasten P and Q close to C A on each side of it, and the bar and the circle will swing as one piece.

Again, if we wish the bar to remain at rest, we may separate P and Q as much as possible, and when the bar has been pushed as far as it can go by one of the pins, it will remain at rest, for the second pin can only just come up to it on the opposite side.

Conceive now that the pin Q is shifted to q, the arm C A will be pushed to Q by the pin P when moving to the right, but will only return as far as Q can carry it, i.e. to q, and the vibration will take place through an angle Q C q instead of an angle Q C P, and in this way, by separating P and Q, the motion of AC may be reduced till it ceases altogether.

So, therefore, we obtain precisely the same result as in the former case, and can advance the ratchet wheel through any integral number of teeth up to a limit fixed by the amount of the vibration of the arm.

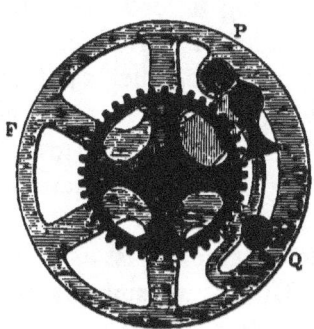

Fig. 102.

The practical arrangement is shown in Fig. 102, where a ratchet wheel, an arm carrying a click, and another wheel provided with a circular slot, are placed in the order stated upon the same axis, and can all move independently of each other. There are two moveable pins in the circular slot, which are capable of being fixed in any position by nuts at the back of the wheel, and which embrace the arm carrying the click, but do not reach the click itself.

The ratchet wheel is connected with a screw which advances the cutter across the table, and the object is to impart definite but varying amounts of rotation to the screw after each cut has been taken.

The wheel F receives a fixed amount of vibration from the table, and will impart the whole thereof to the click if P and Q be made to embrace the arm closely upon each side ; or it will impart any less amount, gradually diminishing to zero, as P and Q are separated to greater intervals along the groove, and thus the feed of the cutter may be regulated.

61. A mechanical equivalent for the teeth and click may be found in what is termed a *nipping lever*, constructed upon the following principle.

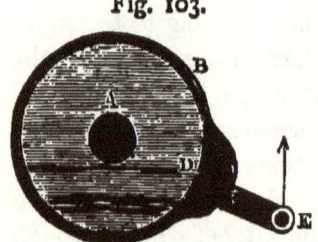

Fig. 103.

Conceive that a loose ring, B, surrounds a disc, A, and that upon a projecting part of the ring there is a short lever, D E, centred. This lever is moveable about a fulcrum at F, near to the wheel, and terminates at one end in a concave cheek, D, fitting the rim of the disc. On applying a force at E the lever will nip or bite upon the disc, and the friction set up may be enough to cause them to move together as if they were one piece. This friction does, in fact, increase almost indefinitely, for the harder you pull at E the greater will be the pressure at D, and since the friction increases with the pressure, being always proportional to it in a fixed ratio, the force of friction will be developed just as it is wanted.

Upon reversing the pressure at E, the nipping lever will be released and the ring will slide a short space upon the disc : thus the action of a ratchet wheel is imitated.

Silent Feed.

62. The ratchet wheel has been much used in obtaining an advance of the piece of timber at each stroke of the saw in sawing machines. A substitute has been found in an adaptation of this nipping lever, and is commonly known as the *silent feed*.

An arm A B (Fig. 104), centred at C, rides upon a saddle which rests upon the outer rim of a wheel; a piece, E E, is attached to one end of the arm, and admits of being pressed firmly against the inside of the rim of the wheel which carries the saddle.

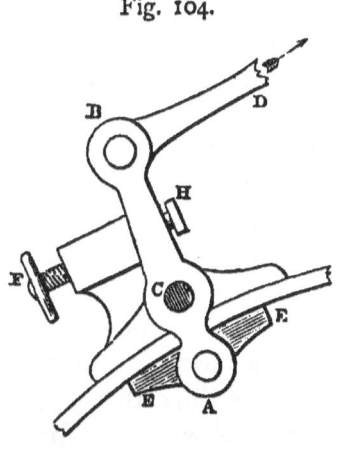

Fig. 104.

It is clear that when the end, B, of the arm, A C B, is pulled to the right hand, the rim of the wheel will be grasped or nipped firmly between the saddle and the piece, E E, and that the pull in B D will move the saddle and wheel together, as if they were made in one piece. When B is pushed back, a stop prevents B C A from turning more than is sufficient to loosen the hold of E E, and the saddle slides upon the rim through a small space.

In this way the action of a ratchet wheel is arrived at, and, by properly regulating the amount of motion communicated by the link B D, we obtain an equivalent for a ratchet wheel with an indefinite number of teeth.

It is this circumstance which renders the contrivance so useful. The amount of feed of the timber can be regulated with the utmost nicety; the opposite end of the link B D is moveable by a screw, according to the principle laid down in Art. 59, and can be set at any distance from the centre of the arm upon which it rides, and thus the contrivance is as perfect as can well be imagined.

A screw, F, may be employed to bring up a stop, H, towards the arm, A C B, and so to prevent the arm from twisting into the position which gives rise to the grip of E E. The saddle will then slide in both directions without imparting any motion to the wheel, a result which is obtained in an ordinary ratchet wheel by throwing the click off the teeth.

63. Where the ratchet wheel moves at *each* vibration of the driver, and not during every alternate movement, an *escapement*, or something approximating thereto, must be employed. The action now takes place alternately upon opposite sides of the wheel.

Upon looking back to the elementary form of escapement described in Art. 14, it is quite apparent that if the frame A B be moved to the left, the pallet D will push P before it, sufficiently far to bring R in front of C, and then, upon the return of the frame, the pallet C will push R before it, and thus the scape wheel will rotate always in one direction.

We remark that this direction is the *opposite* to that in which the wheel revolves when driving the escapement.

The same thing is true, generally, of all *Recoil Escapements*, and upon examining them it will be found that the scape wheel may be driven backwards by the pallets.

64. A like action results where two clicks are hung upon a vibrating bar and one of them terminates in a hook.

Fig. 105.

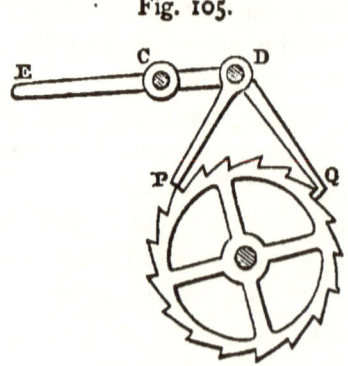

The bar E C D vibrates on C as a centre, and the pieces Q D, P D, hang at the extremity D. (Fig. 105.)

When P pushes on the wheel, the arms P D and D Q open, and the hook at Q slips over a tooth ; whereas, upon reversing the motion of the driving lever, the hook drags the wheel with it, and P slips over a tooth; and thus the wheel advances upon each vibration of the moving arm.

This contrivance is due to Lagarousse.

The click may be replaced by a hook turned in the reverse direction, as in the annexed example, which is taken from spinning machinery.

Here, the bell crank lever E C D is furnished with the hooks B P, D Q, and by swinging it to and fro through the angle E C e, we shall catch up a tooth at the points P and Q alternately, and shall drive the wheel in one direction.

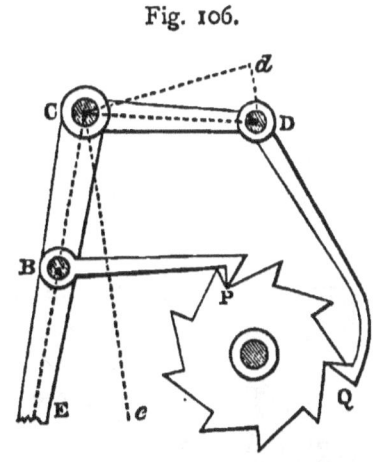

Fig. 106.

The hooks may both be replaced by clicks, turned in the reverse direction, as in the annexed example, which is taken from the specification of a patent planing machine, where it was intended that the cutter should act during each movement of the table.

Fig. 107.

Here the levers which carry the clicks are moveable about centres at A and B, and are connected by a link, E F. The two arms A E and F B vibrate together, and the action is

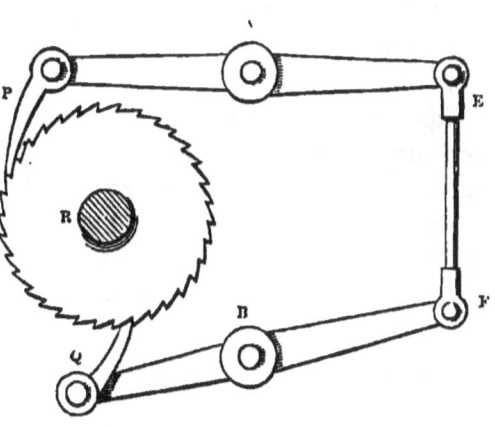

precisely the same as that which we have been considering.

H 2

65. The converse of the contrivance described in Art. 24 may also be used to convert reciprocating into circular motion; that is to say, the bar may be employed to turn the screw barrel, instead of the screw driving the bar; but such an arrangement gives rise to a great increase of friction, and is only met with when a small amount of force is to be exerted.

There is a well-known instance in light hand drills, called Archimedean drills, where the rotation is derived by pushing a nut up and down a rod, upon which a screw of rapid pitch is formed, the drill rotating in opposite directions as the nut rises and falls.

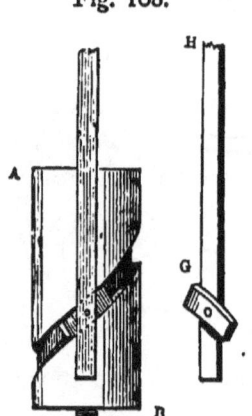

Fig. 108.

This movement was at one time proposed by Sir J. Whitworth in order to obtain a reversing motion of the cutter in planing machinery.

Here a rod, H G, was provided with a sort of tooth, G, which fitted exactly into a groove in the form of a screw-thread traced upon the cylinder A B. As the rod moved up and down it reversed the position of the cutter, and enabled it to act while the table was moving in either direction.

This reversing motion has not been used, but another has been employed in the place of it, where an endless catgut band which runs round a pulley capable of vibrating through a given angle, is carried on to a small pulley attached to the top of the tool box, the cutter turning through two right angles when the band is made to traverse a small space in opposite directions by the action of the vibrating pulley.

This band is kept stretched by a tightening pulley, and a second turn round one or more pulleys in the circuit will always prevent the band from slipping.

CHAPTER III.

ON THE TEETH OF WHEELS.

66. We propose now to enter into a mathematical investigation of the forms of teeth adapted for the transmission of motion or force in combinations of wheelwork, and we have already stated the general nature of the problem.

It is required to shape the teeth or projections upon the edges of two circular discs in such a manner that the motion resulting from the mutual action of the teeth upon the discs or wheels shall be precisely the same as the rolling action of those definite circles known as the pitch circles of the wheels in question.

The various geometrical propositions which enable us to accomplish the solution of the problem will be given in detail, and the practical application of these theorems will also be indicated as briefly as possible.

Commencing, then, with our circular discs, which, in the case of light wheels, such as are used in clockwork, would be castings in brass or gun metal, having light rims of sufficient substance to allow of the cutting away of the material so as to shape the teeth, we should settle in the first instance the exact size of the pitch circles, and next the number and pitch of the teeth which we meant to employ.

The last enquiry involves only a very elementary knowledge of geometry: we have merely to find out how many times we can repeat the space occupied by the pitch of a tooth upon a circle of given diameter, so as quite to fill up the circumference without any error in excess or defect.

67. It will be seen at once that the following very simple equation connects the diameter of a pitch circle with the number of teeth and their pitch.

Let D be the diameter of the pitch circle in inches,
P the pitch of a tooth *in inches*,
n the number of teeth upon the wheel.

Then $n \cdot P =$ circumference of pitch circle $= \pi D$,
where $\pi = 3\cdot 14159$, or $= \tfrac{22}{7}$ approximately.

$$\text{Hence } n = \frac{\pi}{P} \cdot D$$

$$\text{or} \quad D = \frac{P}{\pi} \cdot n.$$

In order to save trouble, definite values are assigned to P, such as $\tfrac{1}{4}, \tfrac{3}{8}, \tfrac{1}{2}, \tfrac{5}{8}, \tfrac{3}{4}, \tfrac{7}{8}, 1, 1\tfrac{1}{8}, 1\tfrac{1}{4}, 1\tfrac{1}{2}, 2, 2\tfrac{1}{2}$, &c., and the values of $\frac{\pi}{P}, \frac{P}{\pi}$, are calculated and registered in a table of which we give a specimen.

	$\frac{\pi}{P}$	$\frac{P}{\pi}$
2	1·5708	·6366
2¼	1·3963	·7135
2½	1·2566	·7958
2¾	1·1333	·8754
3	1·0472	·9548

Thus, let P $= 2\tfrac{1}{2}$ inches, we find in the table that

$$\frac{\pi}{P} = 1\cdot 2566 \text{ and } \frac{P}{\pi} = \cdot 7958.$$

Suppose a wheel of 88 teeth, and $2\tfrac{1}{2}$ inch pitch, to be in course of construction, and that we require to know the diameter of the pitch circle.

$$D = \frac{P}{\pi} \cdot n$$
$$= \cdot 7958 \times 88 = 70\cdot 03 \text{ inches.}$$

Teeth of Wheels.

Or, again, if the diameter of the pitch circle be 70 inches, and the number of teeth of 2½ inch pitch be required,

$$n = \frac{\pi}{p} \cdot D$$
$$= 1\cdot2566 \times 70 = 87\cdot96$$
$$= 88 \text{ very nearly.}$$

In practice it is more easy to treat of the subdivision of a straight line than of the circumference of a circle, and it is the custom therefore to suppose the diameter of the pitch circle to be divided into as many equal parts as there are teeth upon the wheel, and to designate $\frac{D}{n}$ as the diametral pitch in contradistinction to P, the circular pitch.

Further, let $\frac{D}{n} = \frac{1}{m}$ where m is a whole number,

now $\frac{D}{n} = \frac{P}{\pi} \therefore \frac{1}{m} = \frac{P}{\pi}$, or $P = \frac{\pi}{m}$.

The values of m and $\frac{\pi}{m}$ are registered in a table, of which a portion is given, and the circular pitch can be at once found.

Values of m	3	4	5	6	7	8	9	10	12
Values of P	1·047	·785	·628	·524	·449	·393	·349	·314	·262
or approximately	1	¾	⅝	½	$\frac{7}{16}$	⅜	$\frac{11}{32}$	$\frac{5}{16}$	¼

Thus, let D = 8 inches, n = 80,

$$\therefore \frac{D}{n} = \frac{8}{80} = \frac{1}{10}$$

Hence P = ·314 = $\frac{5}{16}$ inch.

68. A very simple piece of apparatus, called a *sector*, is constructed to save workmen trouble in arranging the sizes of wheels with given numbers of teeth to work with given pinions or conversely.

It consists of two light straight arms of brass, centred upon a pivot, with an arc of a circle at the end of one of the arms, and a binding screw to clamp the other arm to the arc. Thus the arms can be set at any angle. These bars are *each* graduated at intervals along their length, the intervals being equal except for some of the smaller numbers up to 10, and the reading may extend to 150 or thereabouts.

Conceive now that a pinion with 12 leaves or teeth is placed between the bars, and that they are closed up till the pinion just reaches the graduation 12, and let it be required to find the size of a wheel having 96 teeth to work with the pinion. We should measure the distance across the bars at the graduations 96, and that would give the required diameter.

This is evident, because if a disc were placed between the bars with its plane perpendicular to the plane of the bars, its diameter would form the base of an isosceles triangle, of which the two legs form the sides, and the bases of any of these triangles must be in the same proportion as the respective sides so long as the vertical angle remains unchanged.

Thus the readings of the legs give the proportionate diameters of the wheels and pinions, and hence also the proportionate numbers of teeth in the same.

As far as we have gone at present, we have simply determined the relation between the pitch of a tooth and the circle upon which it is formed; it will be necessary to consider also how much of this arc called the pitch is to be occupied by the solid tooth, and how much by the adjoining empty space; we must arrange also the depth of the tooth and the depth of the open space. But without touching upon these points it will be convenient to work out this problem of shaping the teeth in such a manner that the wheels shall roll with perfect accuracy upon each other, precisely as the ideal pitch circles, which we have already referred to, would move by rolling contact.

The first proposition is almost self-evident.

69. *When two circles roll together, their angular velocities are inversely as the radii of the circles.*

Let the circles centred at A and B move by rolling through the corresponding angles P B D and Q A D. (Fig. 109.)

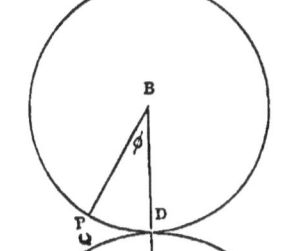

Fig. 109.

Let $\left. \begin{array}{l} AD = a \\ BD = b \end{array} \right\} \left. \begin{array}{l} QAD = \theta \\ PBD = \varphi \end{array} \right\}$

Then $QD = a\theta$, $PD = b\varphi$,

But $QD = PD$

$\therefore a\theta = b\varphi$

$\therefore \dfrac{\theta}{\varphi} = \dfrac{b}{a}.$

But the angular velocities of the circles are as the angles described by each of them in the same given time,

$\therefore \dfrac{\text{angular vel. of A}}{\text{angular vel. of B}} = \dfrac{b}{a} = \dfrac{BD}{AD}$

which proves the proposition.

70. The next proposition is a corollary to Art. 53.

When two arms or cranks are connected by a straight link, the angular velocities of the arms will be inversely proportional to the segments into which the direction of the link divides the line of centres.

The proof already given (*see Art.* 53) must be carried a step further by connecting the ratio of the perpendiculars upon the link with that of the segments of the line of centres, and is, when so extended, sufficiently simple to be understood by everyone, but we add another investigation which may be satisfactory to those who have studied mathematics.

If the reader is unable to comprehend what follows, he should merely take the figure as representing the two arms and the link, and he will see that upon producing Q P to meet B C in E he can infer the relation

$$\frac{\text{angular vel. of BQ}}{\text{angular vel. of CP}} = \frac{CS}{BR} = \frac{CE}{BE}.$$

Conceive that CP and BQ represent two arms, centred at B and C respectively, and connected by a link PQ.

Let the line of direction of QP meet BC in E, draw CS and BR perpendicular to PQ or PQ produced,

and let $\left. \begin{array}{l} CP = a \\ BQ = b \end{array} \right\} \left. \begin{array}{l} CS = p \\ BR = q \end{array} \right\} \left. \begin{array}{l} PQ = d \\ CB = c \end{array} \right\}$,

θ, ϕ, ψ being the angles marked in the sketch.

Fig. 110.

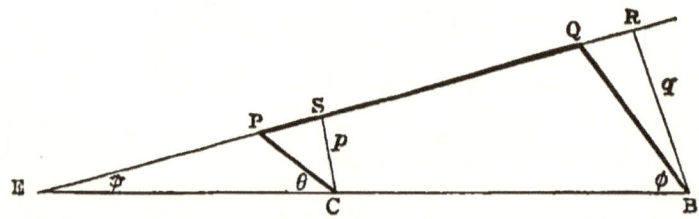

Then $p = a \cdot \sin(\theta + \psi)$
$q = b \cdot \sin(\phi + \psi)$
$d = a \cos(\theta + \psi) + c \cos \psi - b \cos(\phi + \psi)$

∴ by differentiation we have

$0 = -a \sin(\theta + \psi)(d\theta + d\psi) - c \sin \psi \, d\psi$
$\qquad + b \sin(\phi + \psi)(d\phi + d\psi)$
∴ $c \sin \psi \, d\psi = -p(d\theta + d\psi) + q(d\phi + d\psi)$
$\qquad = (q - p) d\psi - p \, d\theta + q \, d\phi.$
But $q - p = c \sin \psi$
∴ $0 = -p \, d\theta + q \, d\phi$
∴ $\dfrac{d\theta}{d\phi} = \dfrac{q}{p} = \dfrac{BR}{CS}$

∴ $\dfrac{\text{angular vel. of CP}}{\text{angular vel. of BQ}} = \dfrac{BR}{CS} = \dfrac{BE}{CE},$

which proves the proposition in its general form.

71. This proposition applies equally when P Q meets A B between the centres A and B, as in the case now to be dealt with, and we must remind the student that geometers have shown that every curved line may at each point be supposed to possess the exact amount of curvature which would be found in a circle of definite size called the *circle of curvature* of the point in question. In the modern editions of Newton's 'Principia' the circle of curvature at any point of a curve is defined as being *that circle which has the same tangent and curvature as the curve has at the point in question.*

It is also capable of proof that no other circle can be drawn whose circumference lies between the curve and its circle of curvature, starting from the point considered.

In order to find this circle of curvature at any point P of a curve, we first draw the tangent at P, we then take a very small arc P Q of the curve terminating in P, and from the other extremity Q of the arc we draw Q R perpendicular to the tangent at P, and meeting it in the point R, the diameter of the circle of curvature will then be the limiting value of the ratio $\frac{(\text{arc P Q})^2}{\text{Q R}}$.

If we were to endeavour to obtain the centre of this circle of curvature by geometrical construction we should have to draw a perpendicular, called a *normal*, to the line touching the curve at the point considered, then to draw another perpendicular, or normal, to the tangent at a point very close indeed to the original point, and upon one side of it; these two lines would intersect in the centre of the circle of curvature of the point in question.

Thus the radius of the circle of curvature might be found if our power of execution were equal to the task, but in practice it is only possible to deduce its value by analysis with the aid of the differential calculus, and in treatises on the calculus a whole chapter is devoted to the finding of the circle of curvature of all well-known curves and of curves generally.

72. We must now assume that this circle can be determined, and in Fig. 111 we will make A and B the centres of motion of two pieces provided with teeth of some determinate form, which are in contact at the point p.

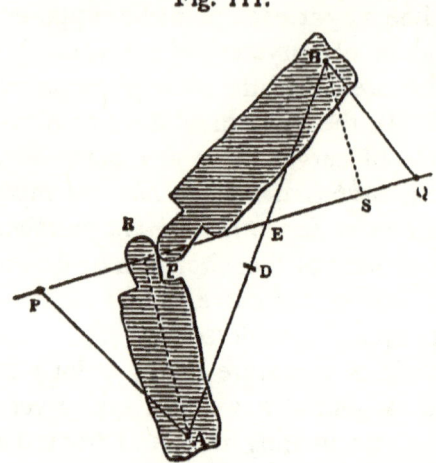

Fig. 111.

Draw now P p Q, a common perpendicular to the surfaces of the teeth at p; and let P and Q be the centres of the circles of curvature of the curves which touch at p; draw also A R, B S, perpendicular to P Q.

Then, in the first instant of motion, P Q *may be regarded as constant,* because it is the distance between the centres of two ascertained circles which do not vary in size for a very small amount of sliding motion of the curves upon each other; and therefore the angular velocities of the two pieces will be identical with those of A P and B Q.

But we have just proved that

$$\frac{\text{angular vel. of A P}}{\text{angular vel. of B Q}} = \frac{BS}{AR} = \frac{BE}{AE}$$

whence $\dfrac{\text{angular vel. of piece A}}{\text{angular vel. of piece B}} = \dfrac{BE}{AE}.$

In order to connect our investigation with this case of link-work motion, we have only to remember that an imaginary combination of the two arms B Q, A P, connected by a link P Q is always supplied, and that although the separate parts of this combination are continually changing, yet it is always present as a whole, and gives us the means of comparing the angular velocities of the pieces A p and B p at every instant.

Suppose it to be required that the angular velocities of the

Teeth of Wheels. 109

two pieces shall be the same as those of the pitch circles of radii A D, B D, which is the case in wheelwork.

We must now form the curves so that E shall coincide with D, and shall never leave it during the motion; in other words, <u>the common perpendicular to the surfaces of any two teeth in contact must always pass through the point of contact of the two pitch circles.</u>

If the teeth can be formed so as to satisfy this condition the problem will be fully solved, and we proceed to give the solutions which have been devised by the ingenuity of mathematicians.

73. There are two curves which will be of great assistance to us, which are the following:

1. An *epicycloid* is a curve traced out by a point, P, in the circumference of one circle, which rolls upon the convex arc of another circle. This curve is represented by the dotted line in Figure 112.

Fig. 112. Fig. 113.

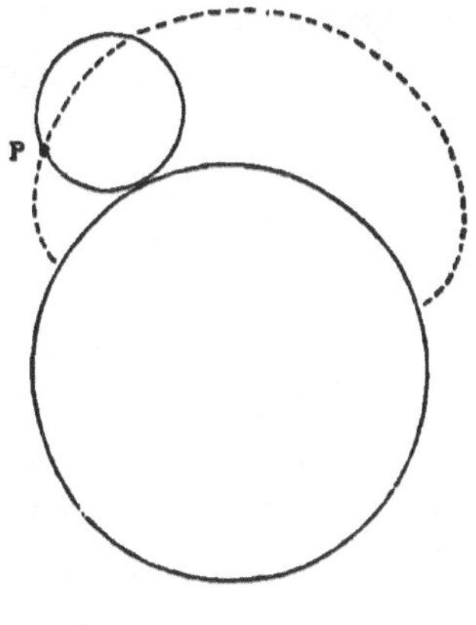

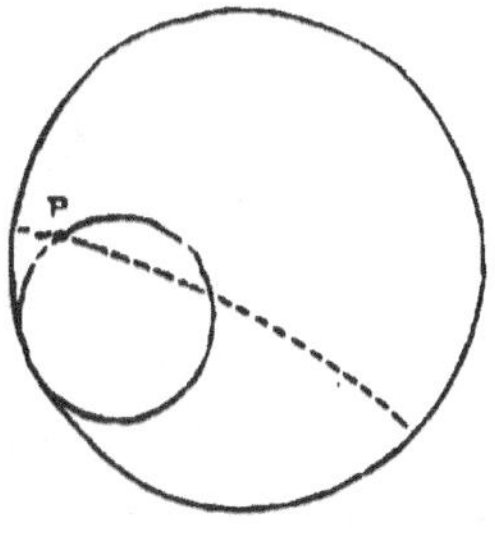

2. A *hypocycloid* is a curve traced out by a point, P, in the circumference of one circle, which rolls upon the concave arc of another circle. This curve is represented by the dotted line in Figure 113.

74. Conceive now that we are about to form these two curves, the one a ~~hypocycloid~~ *epi* upon the outside of the pitch circle A, the other an ~~epi~~ *hypo* *cycloid* upon the inside of the pitch circle B; we will employ the *same* generating circle in both cases, and will suppose that A and B represent the pitch circles of two discs upon which teeth are to be carved out. Our object is to show that teeth shaped according to these curves will answer the purpose.

Having drawn the curves, bring the circles together till the *two circles touch in the line of centres and the two curves in another point* P, as shown, when it will be found that the common perpendicular at the point of contact of these curves passes through D.

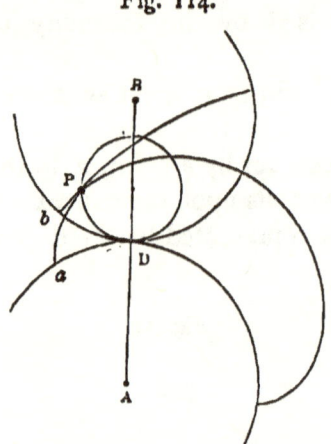

Fig. 114.

The truth of this statement will be evident from the consideration that when the curves touch, the describing circle may be taken as being ready to generate either the one curve or the other; now the describing circle cannot do this unless it be resting upon both circumferences indifferently, that is, resting upon the point where the circumferences of the pitch circles touch each other, in which case the common perpendicular to the curves at P must pass through the point D.

But this is the very condition which we are seeking to fulfil, because if it maintains the teeth will move the discs upon which they are shaped with a relative velocity, which is represented by the ratio $\frac{BD}{AD}$, and we have shown that the relative velocity of the two pitch circles is also $\frac{BD}{AD}$, hence the relative velocity of the discs furnished with teeth is precisely the same as that of the pitch circles, and the problem which we are investigating is solved completely.

Teeth of Wheels.

75. We have now obtained two curves which satisfy the geometrical requirements of the problem, and it remains to put our theory into practice.

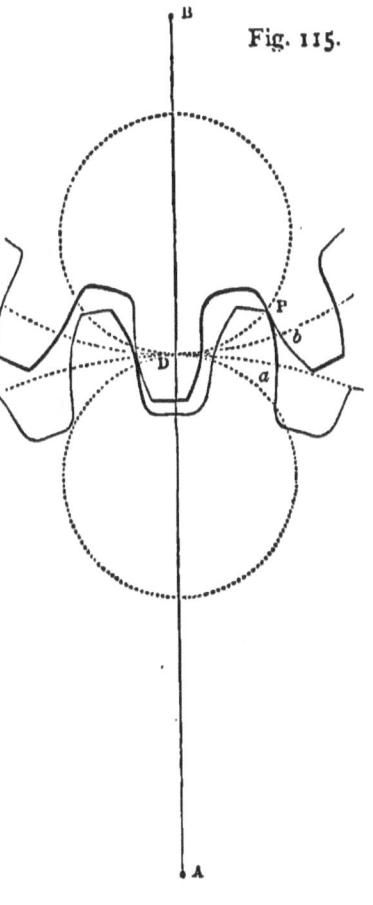

Fig. 115.

The epicycloid and hypocycloid which form the acting surfaces of two teeth must be produced by one and the same describing circle.

Let A and B be the two pitch circles.

Take a circle of any convenient size less than either A or B, as indicated by the small dotted circle, and with it describe an epicycloid upon A and a hypocycloid upon B.

Let these curves determine the acting surfaces aP, bP, of two teeth in contact at P; then the tooth aP will press against bP so that the perpendicular to the surfaces in contact at P shall pass through D, and the relative angular velocities of two pieces centred at A and B, and furnished with these teeth, will be the same as those of the two pitch circles.

As the wheels rotate, we find that the point of contact P travels along the upper small dotted circle starting from D. In the same way the points of contact of teeth to the left of A D B, travel along the lower dotted circle up to D, and it is therefore essential to form the teeth in the manner which we are about to describe by somewhat extending our construction. We have now to make complete teeth upon both wheels, and to provide that either A or B may be the driver.

As far as we have gone we have described the point of a tooth upon A and the flank of one upon B, and have supposed A to drive B. If the conditions were reversed and B were to drive A, we should have to obtain from *one* describing circle the curves suitable for the point of a tooth upon B and the flank of one upon A; this describing circle is not necessarily of the same size as the former one, but it is very advantageous to make it so, and we shall therefore assume that the teeth upon A and B are formed by the *same* describing circle.

Fig. 116.

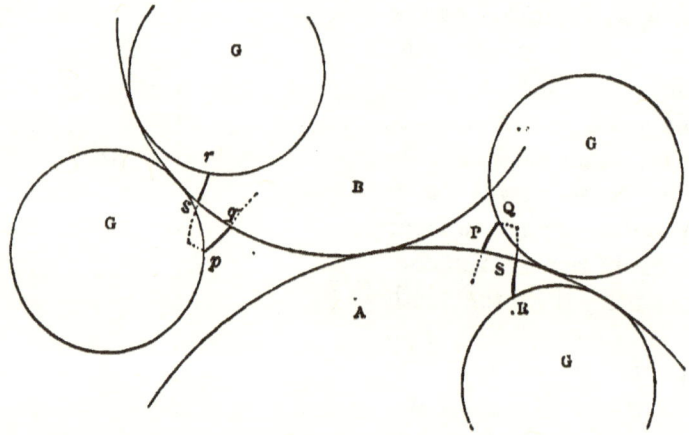

Let the describing circle G trace P Q, S R upon A, and *p q*, *s r* upon B, then the complete teeth can be made up as shown in the diagram.

By preserving a constant describing circle any wheels of a set of more than two will work together, as for example, in the case of change wheels in a lathe.

It remains to trace the changes in the character of the teeth dependent upon changes in the form or configuration of the hypocycloidal portion of the curves.

Teeth of Wheels.

76. If we trace the changes in form of the hypocycloid, as the describing circle increases in size until its diameter

Fig. 117.

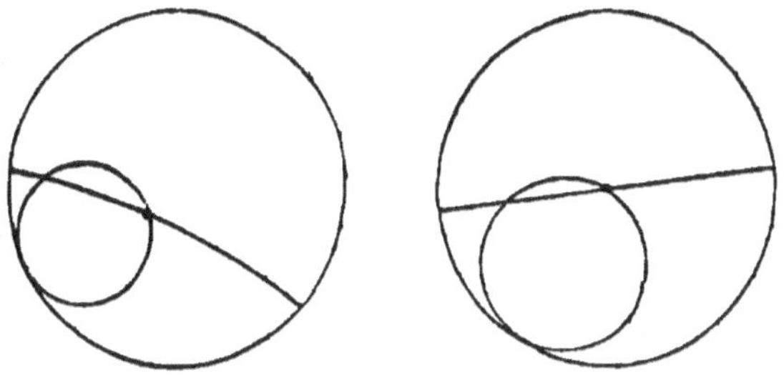

becomes equal to the radius of the circle in which it rolls, we shall find that the curve gradually opens out into a straight line.

It is indeed a well-known geometrical fact that when the diameter of the circle which describes a hypocycloid is made equal to the radius of the circle within which it rolls, the curve becomes a straight line.

Fig. 118.

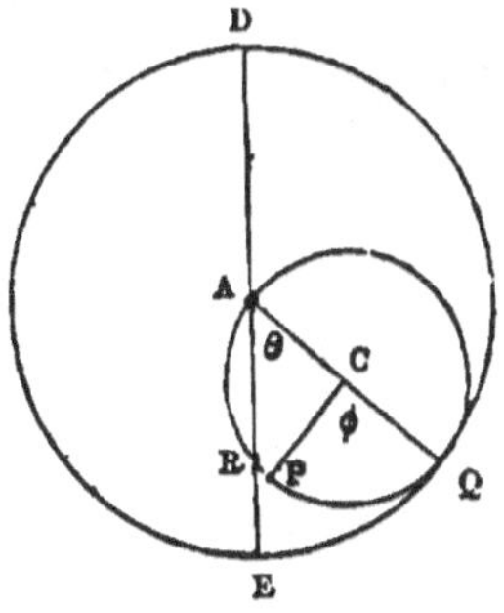

Let C be the centre of the describing circle at any time, and P the corresponding position of the describing point. (Fig. 118.)

Suppose that P begins to move from E, so that the arc P Q shall be equal to the arc E Q.

Join C P, A E; let E A Q $= \theta$, P C Q $= \phi$

    Then   P Q $=$ E Q

    or C Q $\times \phi =$ A E $\times \theta$.

    But   A E $=$ 2 C Q,

∴   C Q $\times \phi =$ 2 C Q $\times \theta$ or $\phi = 2\theta$.

Now ϕ cannot be equal to 2θ unless P coincides with R in the line A E; in which case the diameter E.A D is the path of P.

This property of a hypocycloid is taken advantage of in Wheatstone's *photometer*, where an annular wheel is constructed, and a second wheel of half its diameter is made to run very rapidly upon the internal circumference; a small bead of glass, silvered inside, is attached to a piece of cork fitted on this internal wheel. The bead would give the images of two lights held upon either side of it. When the wheel revolves these small images or spots of light become luminous lines of light, whose brilliancy can be compared, and made equal, by shifting the apparatus towards the weaker light. This contrivance is a philosophical toy, it is not used.

77. The *first* particular case of the general solution is the subject of the present article.

It will be remembered that the hypocycloid determines the flank of the tooth upon either wheel: if, therefore, the radius of the circle describing the hypocycloid be taken in each case to be *half that of the corresponding pitch circle*, the teeth will have straight, or *radial flanks*, as they are commonly called.

The method of setting out the teeth is the following:

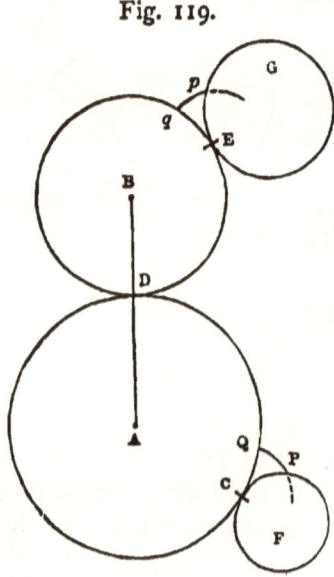

Fig. 119.

Let A and B be the centres of two pitch circles which touch in the point D.

Let a circle, F, *whose diameter is equal to* B D, roll upon the circle, A, and generate the epicycloid Q P; this curve determines the form of the driving surface of the teeth to be placed upon A.

Let another circle, G, *whose diameter is equal to* A D, roll upon the circle B, and generate the epicycloid *q p*; this curve determines the driving surface of the teeth to be placed upon B.

Here of necessity the describing circle is not of the same size when tracing out the points of the teeth upon A and B; but, by reason that the same circle gives the point upon A and the flank upon B, or conversely, and that the flanks in each case are straight lines, the condition in Art. 72 is still fulfilled.

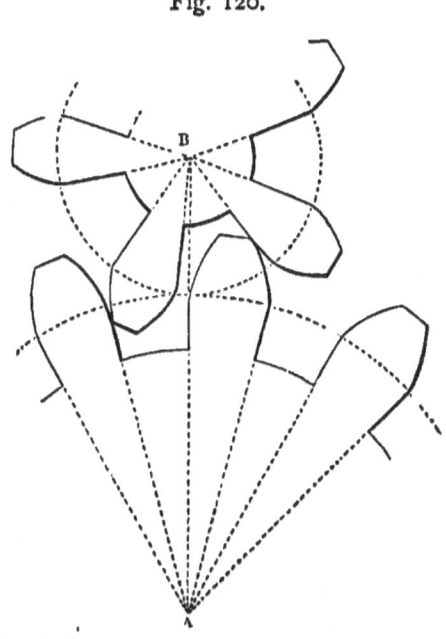

Fig. 120.

The annexed figure shows us these teeth with *radial flanks*, the straight edges of the teeth pointing towards the centres of the respective pitch circles.

78. As the circle describing the hypocycloid goes on increasing until it becomes equal to the circle in which it rolls, the curve passes from a straight line into a curve, and finally degenerates from a small half-loop shown in the sketch down to an actual point.

Fig. 121.

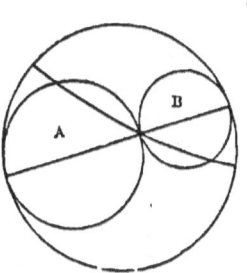 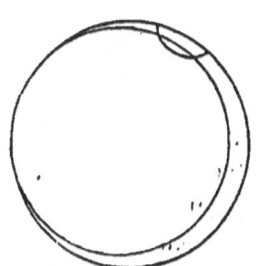

It appears also that the same hypocycloid is generated by

each of the circles A and B, which are so related that the sum of their diameters is equal to the diameter of the circle in which they roll.

79. The *second* particular case of the general solution occurs when the hypocycloid degenerates to a point; we then obtain a wheel with pins in the place of teeth, and derive a form which is extensively used in clockwork.

Fig. 122.

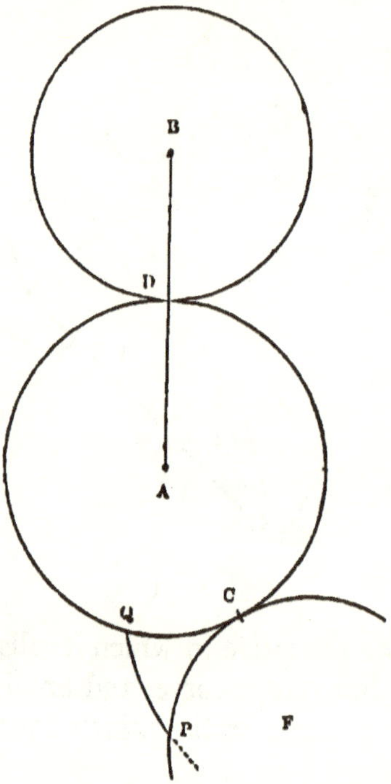

The pin must have some sensible diameter, but we will first suppose it to be a mathematical point.

We have just seen that when the hypocycloid becomes a point, the describing circle must be taken equal to that within which it is supposed to roll.

As before, let A and B be the centres of the two pitch circles touching each other in the point D.

Let a circle, F, equal to B, roll upon the circle A, and generate the epicycloid P Q.

This curve will determine the acting surface of the teeth to be placed upon A, which will work against pins to be placed at intervals on the circumference of the circle B.

Thus we shall have epicycloidal teeth upon the driver, working with hypocycloidal teeth on the follower, but these latter teeth are pins, or mere points theoretically, instead of being curved pieces of definite form. Here it is perfectly apparent that the condition upon which we rely is again fulfilled.

The pin must have some size, and we shall take into

account the size of the pin by supposing a small circle, equal to it in sectional area, to travel along the theoretical path of the point, and to remove a corresponding portion of the curved area occupied by the epicycloids.

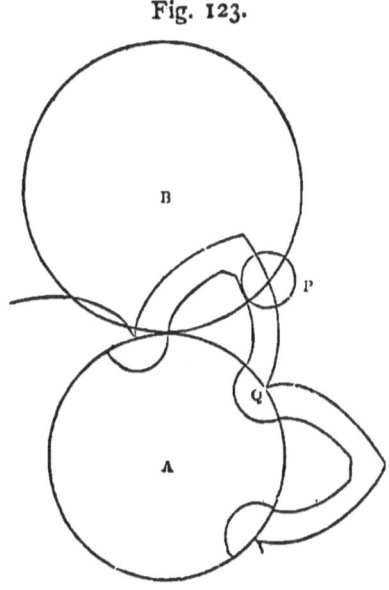

Fig. 123.

Assume that Q P represents the acting surface of a tooth which drives before it a point, P. (Fig. 123.)

Make P the centre of a circle equal to the size of the pin : suppose this circle to travel along P Q, having its centre always in the curve ; remove as much of the tooth as the circle intercepts, and the remainder will give the form of the working portion.

We shall presently find that in practice the pins are always placed upon the *driven wheel*, and as this rule is never broken, for reasons to be stated hereafter, we shall assume it to exist when we come to apply our solution to the case of a rack and pinion.

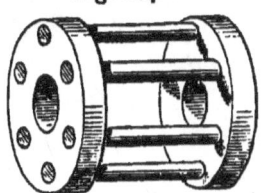

Fig. 124.

There is a very old form of pinwheel, called a lantern pinion, where the pins are made of round and hard steel wire, and are supported between two plates, in the manner shown in the sketch. This form has been much used by clockmakers, because it runs smoothly, and has the merit of combining great strength with durability.

80. If either of the wheels becomes a rack, that is, straightens into the form of a bar, the radius of the pitch circle must be infinitely large; and we shall now take up the inquiry as to the changes introduced into the shape of the teeth by this transition from the circle into a straight line.

The curve which we have called an *epicycloid*, changes into a *cycloid* when the rolling circle runs along a straight line instead of upon the outer circumference of another circle.

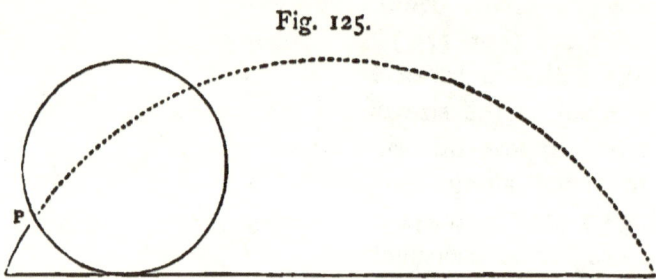

Fig. 125.

It is, in fact, the curve described by a point in the rim of a wheel as it runs along a level road or rail.

It is shown in Fig. 125, and possesses some very interesting properties with reference to the swing of a pendulum; it is therefore a curve very familiar to those who study mechanics.

So far as the general solution in Art. 75 is concerned, the changes will be the following. Conceive that the circle A is enlarged till it becomes a straight line; then the circle G, which rolls upon the inner and outer circumferences of the circle A, tracing thereby the points and flanks of the teeth upon A, will in each case generate the same curve, viz. a *cycloid*.

Thus the teeth upon B will remain as before, and each face of a tooth upon the rack A will be made up of two arcs of cycloids meeting in the pitch line.

81. In Art. 77, where the teeth have *radial flanks*, the matter is not quite so simple, for the describing circles which give the radial flanks are in each case to be of one-half the diameter of the pitch circle in which they respectively roll; and here one of the pitch circles is infinite, whence it follows that a circle half its diameter is infinite also, or may be regarded as a straight line.

The curve traced out by one extremity of a straight line rocking upon the circumference of a given circle, is, of course, the same as that described by one end of a string P Q, which is kept stretched while it is unwound from the circumference of the circle. The end of the line, or the end of the string, is at first at the point A in the curve A P, and the curve is traced out while the line rocks in one direction, or during the unwinding of the string.

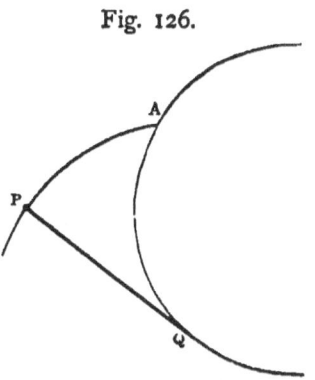

Fig. 126.

This curve A P is a very well-known curve, and is called the *involute* of a circle; we have met with it before in Art. 19 (*note*), and we proceed to show that in the case of a *rack and pinion having teeth with radial flanks*, the driving surfaces of the teeth upon the pinion will be the *involutes of the pitch circle of the pinion* in question.

82. To make this matter clear, we refer to Fig. 127, and observe that the circle F, rolling upon a straight line, generates a cycloid and gives the form of the driving surfaces of the teeth upon the rack: the circle G becomes infinite, and E p changes to a straight line. The change which the rest of the construction undergoes is simply the substitution of the involute $q\,p$ for the corresponding epicycloid, the circle G having passed into a straight line. The change is scarcely visible to the eye, but the form

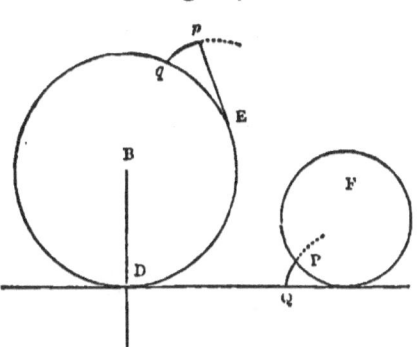

Fig. 127.

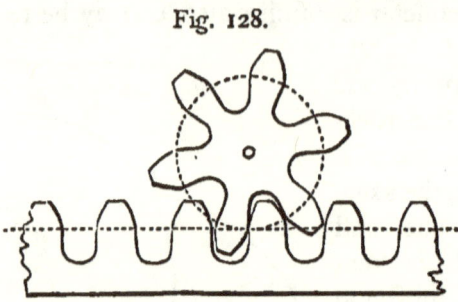

Fig. 128.

of the teeth is shown in the diagram, where the curved portions in the rack are cycloids, the radius of the describing circle being half that of the pitch circle of the pinion, and the curves upon the pinion are the involutes of its own pitch circle.

83. The most perplexing case is the last, where pins are substituted for teeth in either the rack or the pinion, and we construct in accordance with the rule that the *pins are always placed upon the follower*.

1. Let the rack drive the pinion.

Here the circle A becomes infinite, and the curve P Q passes into a cycloid, so that the teeth upon the rack are cycloidal, as shown in Fig. 129.

Fig. 129.   Fig. 130.

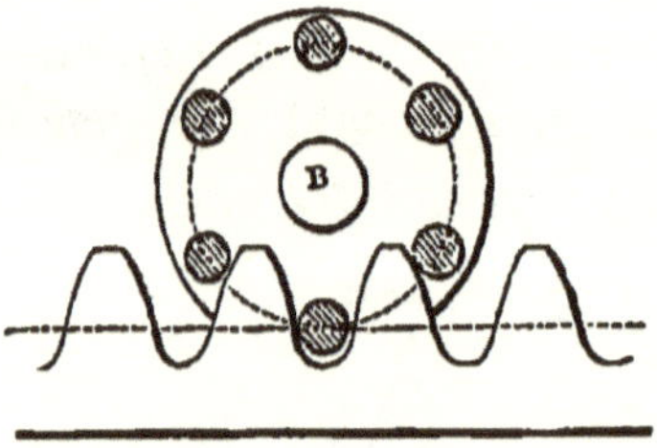

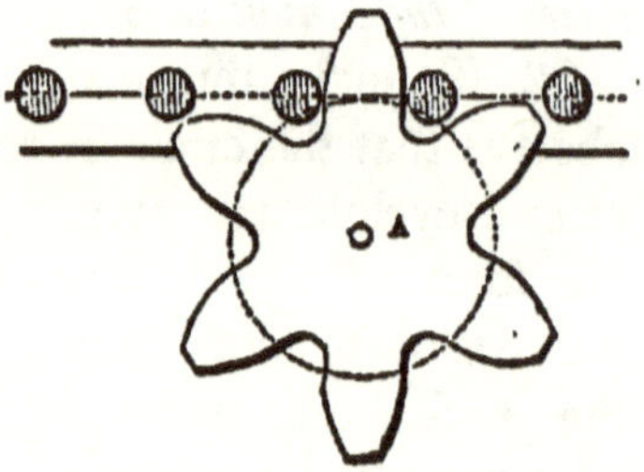

2. Let the pinion drive the rack.

Here the circle B becomes infinitely large, and C P changes into a straight line, the curve P Q passing into the involute of a circle, with the result exhibited in Fig. 130, where the teeth of the driver are the involutes of a circle and are known as *involute teeth*.

84. The last case to be brought before the reader is derived from a property of this involute of a circle, and the teeth are very easily obtained, but are not used in practice, on account of their being unsuited for the transmission of any considerable forces.

We proceed to show that the geometrical requirements of our construction are fulfilled completely by involute curves.

Let A and B represent the centres of two pitch circles touching at the point D, as shown by the dotted lines, and with B as a centre and any line B Q, less than B D, as radius, describe another circle. Through D draw D Q touching this smaller circle, draw A R $\perp^r$ Q D produced, and with centre A and radius A R describe a circle touching Q R in the point R.

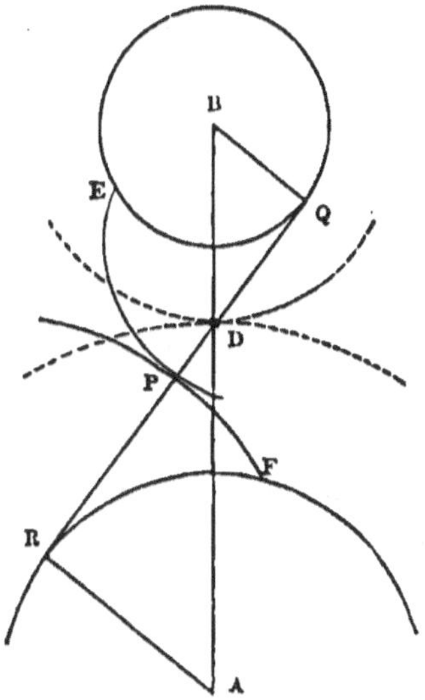

If now we take any point P in Q R, and describe the involutes E P and F P by winding two portions of strings, such as P Q and P R, back again upon their respective circles, we shall have two forms of imaginary teeth in contact, viz. E P and F P, such that

Fig. 131.

(1) These teeth have a common perpendicular to their surfaces at P, viz. R P Q.

(2) This perpendicular cuts the line of centres in a fixed point D.

But these are the conditions which we are seeking to fulfil.

No more direct illustration of our leading proposition could be conceived than this one.

The lines P R and P Q are the respective radii of curvature of the involute curves in contact at P, while R Q, which is equal to R P + P Q, is the link of constant length connecting the arms A R and B Q.

The angular velocities of A R and B Q are therefore as B Q

to A R, or as B D to A D, and this ratio remains constant so long as the curves E P, F P, remain in contact.

85. In order to construct the teeth we must draw our pitch circles touching in D, and then select some angle B D Q at which to draw the line R D Q. When this angle is determined, we obtain the circles of radii B Q, A R, by dropping ⊥^{rs} upon R D Q from the centres A and B, and we then describe teeth of the required pitch by constructing the involutes of these two circles respectively.

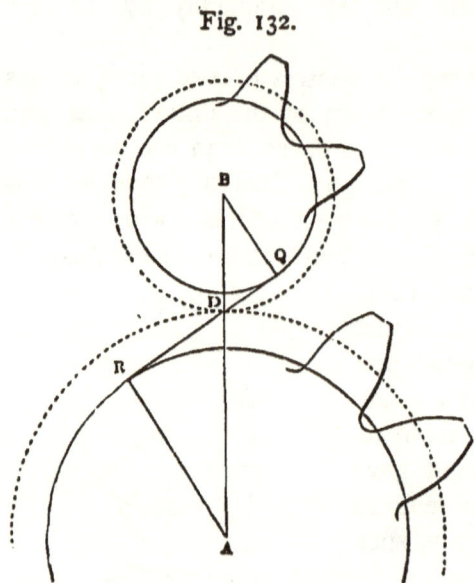

Fig. 132.

We observe, of course, that a great latitude is introduced from the circumstance that A D and B D remain constant while A R and B Q may have different values.

In teeth of this kind there is no difference between the point and the flank: the whole of each edge of a tooth is one and the same curve, viz. the involute of one of the two arbitrary circles.

And further, the points of contact of two teeth must lie either in the line R D Q, or in a second line passing through D, and touching both the circles upon the opposite sides.

86. To adapt this solution to the case of a rack and pinion, we note that one of the circles becomes infinite, and further, that the involute of the infinite circle of radius A R is a straight line perpendicular to its circumference, or perpendicular to Q D. Hence the teeth of the rack are straight lines perpendicular to the direction of Q D.

The direction of D Q is arbitrary; but when it has once been assumed, the radius B Q will be determined, and involute teeth can be formed upon B, the teeth of the rack being straight lines inclined to the pitch line at an angle equal to B D Q.

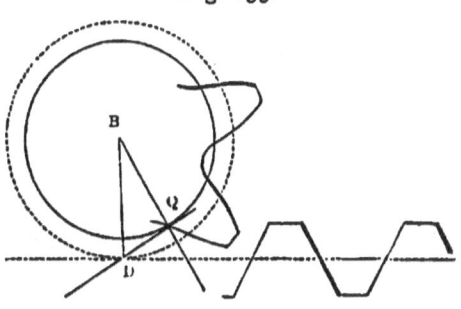

Fig. 133.

87. There are now sundry general points for consideration. We may inquire, where does the action of two teeth begin, and where does it leave off?

Referring to the solution in Art. 75, we observe that if the motion takes place in the direction of the arrows, and the describing circle be placed so as to touch either pitch circle in D, the contact of two teeth commences somewhere in aD, travels along the arc a D b, and ceases somewhere in D b.

Since a D lies entirely without the pitch circle B, it is clear that the action in a D is due solely to the fact that the teeth upon B project beyond the pitch circle B, and similarly that the action in D b depends upon the projections or *points* of the teeth upon A.

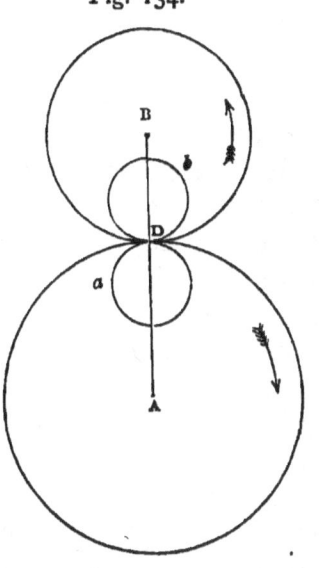

Fig. 134.

It is further evident that the greater the number of teeth upon the wheels, the closer is their resemblance to the original pitch circles, and the more nearly is their action confined to the neighbourhood of the point D.

By properly adjusting the amount to which the teeth are allowed to project beyond the pitch circles, and also their numbers, we can assign any given proportion between the arcs of contact of the teeth upon either side of the line A D B.

Where the teeth upon B are pins, there is comparatively very little action before the line of centres, and there would be none at all if the pins could be reduced to mere points, as in that case there would be nothing projecting beyond the pitch circle B.

Again, since the line D P in Art. 74 is a perpendicular to the surfaces in contact at P, it follows that the more nearly D P remains perpendicular to A D B, the less will be the loss of the force transmitted between the wheels.

Here we have an additional reason for keeping the arc of contact as close as possible to the point D; there is a sensible loss of power as soon as the line D P differs appreciably from the direction perpendicular to A B.

It is on this account that involute teeth are not used in machinery calculated to transmit great force: the line R P D Q in Art. 84 is always inclined to the line A D B at a sensible angle, and a direct and useless strain upon the bearings of the wheels is the result.

88. In combinations of wheel-work, the accurate position of the centres must be strictly preserved; all the solutions given above, with one exception, entirely fail if there be any error in centring the wheels; they are totally vitiated if anything arises to deprive them of their geometrical accuracy. The exception occurs with involute teeth: the position of the centres determines the sum of the radii of the pitch circles, and the wheels will work accurately as long as the teeth are in contact at all.

We see too that teeth with radial flanks are not suitable for a set of change wheels; the describing circles of one pair of wheels are derived directly from their pitch circles, and cannot be adapted to any other pair in the series.

Teeth of Wheels.

Where, however, the solution in Art. 75 is employed, the describing circles *may* be made the same for all the pitch circles, instead of varying with each one of the series, and in that case any pair of wheels will work truly together.

89. As regards the strength of the teeth, we remark that this quality is influenced by the size of the describing circle.

If the diameter of the describing circle be less than, equal to, or greater than the radius of the pitch circle, we shall have the flanks as shown in the sections a, b, c of the sketch.

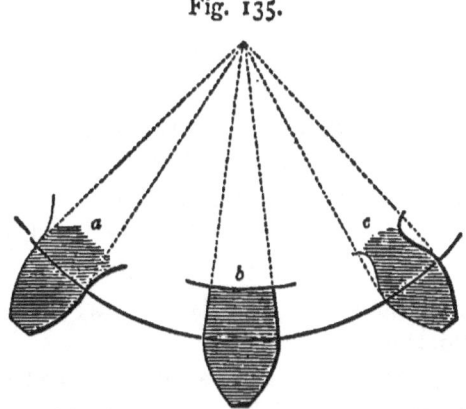

Fig. 135.

It is evident that a small describing circle makes the teeth strong, and that it would be unwise to have them weaker than they are with radial flanks. The form of involute teeth being somewhat similar to that of a wedge, the teeth of this character are usually abundantly strong.

90. It will be proved when we treat of rolling curves that the surface of one tooth must always slide upon that of another in contact with it, except at the moment when the point of contact is passing the line of centres.

This matter should be well understood, the teeth are perpetually rubbing and grinding against each other; we cannot prevent their doing so; our rules only enable us so to shape the acting surfaces that the pitch circles shall roll upon each other.

Nothing has been said about the teeth rolling upon each other ; it is the pitch circles that roll ; the teeth themselves slide and rub during every part of the action which takes place out of the line of centres.

Since, then, the friction of the teeth is unavoidable, it only remains to reduce it as much as possible, which will be effected by keeping the arc of action of two teeth within reasonable limits.

Generally, the friction before a tooth passes the line of centres, is more injurious than that which occurs after the tooth had passed the same line : the difference between drawing a walking-stick along the ground after you and pushing it before you, is given by Mr. Denison as an illustration of the difference between the friction before and after the line of centres ; but this difference is less appreciable when the arc of contact is not excessive.

Where a wheel drives another furnished with pins instead of teeth, the friction nearly all occurs after the line of centres ; hence such pin wheels are very suitable for the pinions in clockwork.

91. When the axes are not parallel we must employ *bevil wheels*, the teeth upon which are formed by a method due to Tredgold.

Let F E D H, K E D L, represent the frusta of two right cones, whose axes meet in C, and which are therefore capable of rolling upon each other.

Let it be required to construct teeth upon two bevil wheels which shall move each other just as these cones move by rolling contact.

Draw A D B perpendicular to D E, meeting the axes of the cones in the points A and B.

Suppose the conical surfaces, H A D, B D L, to have a real existence, and to be flattened out into the circular segments D R, D S ; these segments will roll upon each other just as the circular base H D rolls upon the circular base D L.

Hence these segments will serve as pitch circles, upon

which teeth may be constructed by the previous rules : such teeth may be formed upon a thin strip of metal, and their outline can then be traced upon the surface of the cone terminating in A.

Fig. 136.

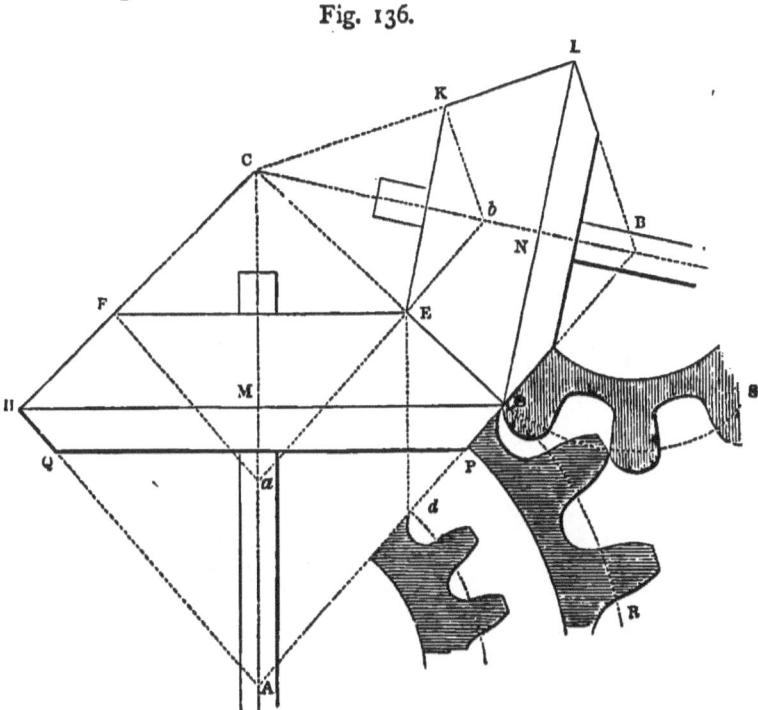

Similarly, if $b \, \mathrm{E} \, a$ be drawn perpendicular to E D, the circle of radius $\mathrm{A} \, d = \mathrm{E} \, a$ will be the pitch circle for the teeth upon the conical surface E a F; the teeth will taper from D to E, and the intermediate form will be determined by a straight line moving parallel to itself, and originally passing through the points D and E.

92. Two very simple questions relating to the transfer of motion by wheel-work remain to be determined.

1. Let two axes be parallel, and let $\dfrac{m}{n}$ be the velocity ratio to be communicated between them.

If a be the distance between the axes, and r, r', be the radii of the two pitch circles A and B.

The condition of rolling gives us $\dfrac{r}{r'} = \dfrac{n}{m} = \dfrac{\text{vel. of B}}{\text{vel. of A}}$.

Also $r + r' = a$, $\therefore r + \dfrac{rm}{n} = a$,

$\therefore r = \dfrac{na}{m+n}$, and $r' = \dfrac{ma}{m+n}$;

whence r and r' are known in terms of m, n, and a.

2. Let the axes meet in a point, and let it be required to construct two *cones* which shall communicate the same velocity ratio by rolling contact.

We now refer to Fig. 136, and assume D N, D M to be the radii of the bases of the two cones L C D, H C D, whose angular velocities are as the numbers m and n respectively.

Let M C N = a, N C D = θ,
Then D N = C D sin θ,
D M = C D sin $(a - \theta)$;
$\therefore \dfrac{\text{D M}}{\text{D N}} = \dfrac{\sin(a-\theta)}{\sin \theta}$.

But $\dfrac{\text{D M}}{\text{D N}} = \dfrac{m}{n}$, since the inverse ratio of the radii of the bases of the cones is the velocity ratio between the axes,

$\therefore \dfrac{m}{n} = \dfrac{\sin(a-\theta)}{\sin \theta} = \sin a \cot \theta - \cos a$,

whence $\tan \theta = \dfrac{n \sin a}{m + n \cos a}$.

If ϕ be an angle such that $n \cos a = m \cos \phi$, $(m > n)$

we have $\tan \theta = \dfrac{n \sin a}{m(1 + \cos \phi)} = \dfrac{n \sin a}{2m \cos^2 \frac{\phi}{2}}$,

whence θ is expressed in a form adapted for logarithmic computation.

Cor.—If $a = 90$, we have $\tan \theta = \dfrac{n}{m}$.

CHAPTER IV.

ON THE USES OF WHEELS IN TRAINS.

93. When a train of wheels is employed in mechanism, the usual arrangement is to fasten two wheels of unequal size upon every axis except the first and last, and to make the larger wheel of any pair gear with the next smaller one in the series.

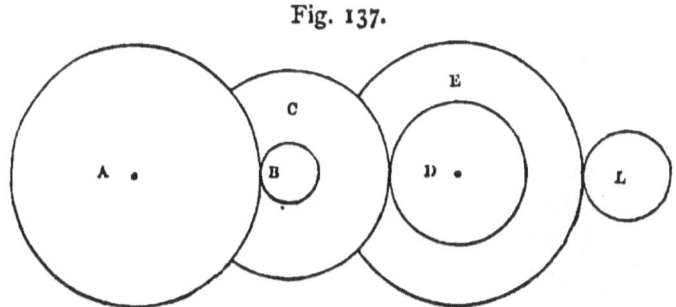

Fig. 137.

Let A be the driver, L the extreme follower, and conceive that L makes (e) revolutions while A makes one revolution;

then $e = \dfrac{\text{number of revolutions of L in a given time}}{\text{number of revolutions of A in the same time}}$.

It will be convenient to distinguish (e) as the *value of the train*, and the ratio which it represents may be at once found when the numbers of teeth upon the respective wheels are ascertained.

Suppose that A, B, C, D, &c., represent the numbers of teeth upon the respective wheels, thus we infer from the condition of rolling that

$$\dfrac{\text{number of revolutions of B in a given time}}{\text{number of revolutions of A in the same time}} = \dfrac{A}{B};$$

and similarly for each pair of wheels:

$$\therefore e = \frac{A}{B} \times \frac{C}{D} \times \frac{E}{F} \times \&c., \ldots \frac{K}{L}.$$

It may frequently simplify our results if we regard e as positive or negative according as A and L revolve in the same or in opposite directions: thus, in a train of two axes, e would be negative, and in a train of three axes it would be positive.

The comparative rotation of wheels is estimated in various ways, thus:

Let N, n be the numbers of teeth upon two wheels.

R, r their radii.

P, p their periods of revolution.

X, x the number of revolutions made by each wheel in the same given time.

It is easy to see that

$$e = \frac{N}{n} = \frac{R}{r} = \frac{P}{p} = \frac{x}{X}.$$

Note.—A belt and a pair of pulleys supply a mechanical equivalent for two working wheels: the belt may be open or crossed, and in either case

$$\frac{\text{the number of revolutions of B in a given time}}{\text{the number of revolutions of A in the same time}} = \frac{\text{diameter of A}}{\text{diameter of B}}.$$

The crossing of the belt merely reverses the direction of one of the pulleys.

Ex. Suppose that we have a train of five axes, and that

1. A wheel of 96 drives a pinion of 8.

2. The second axis makes a revolution in 12 seconds, and the third axis in 5 seconds.

3. The third axis drives the fourth by a belt and a pair of pulleys of radii 20 and 6 inches.

4. The fourth axis goes round twice while the fifth goes round three times.

$$\text{Here } e = \frac{96}{8} \times \frac{12}{5} \times \frac{20}{6} \times \frac{3}{2} = 144$$

or the last axis makes 144 revolutions while the first axis goes round once.

94. It is very obvious that a wheel and pinion upon the same axis is a combination equivalent to a lever with unequal arms, and modifies the force which may be transmitted through it, and, further, that a single wheel is equivalent to a lever with equal arms, and produces no modification in the force which may pass through it.

So, therefore, when any number of separate wheels are in gear, no two wheels being upon the same axis, they are equivalent to a single pair of wheels, viz. the first A, and the last L; the intermediate wheels act as carriers only, and transfer the motion through the intervening space.

Fig. 138.

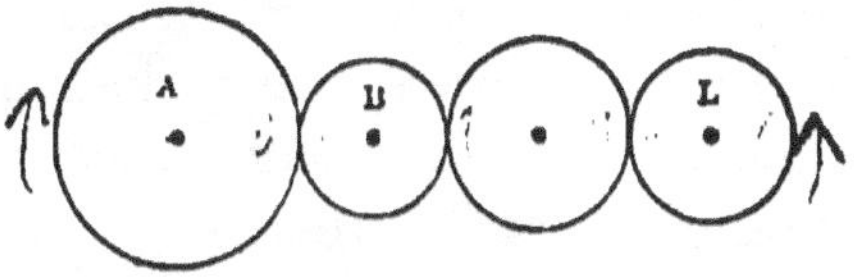

This also appears from the formula, where we find that

$$e = \frac{A}{B} \times \frac{B}{C} \times \frac{C}{D} \times \ldots \frac{K}{L}$$
$$= \frac{A}{L},$$

which is the same result as if A and L were alone concerned in the movement.

95. If, however, a single wheel, such as B, be interposed between two other wheels A and L, although B will not modify the force transmitted, nor alter the velocity, it may be useful in changing the direction in which the wheel L would otherwise revolve. An intermediate wheel so introduced is technically called an *idle wheel*, and we give an instance where this intermediate wheel serves a very useful purpose in causing two other wheels to rotate in the same direction with precisely the same velocity.

The Blanchard turning-lathe, of which a portion is shown

132 *Elements of Mechanism.*

in the sketch, is used for shaping the spokes of wheels, gunstocks, shoe-lasts, and other pieces of an arbitrary form, which no one could imagine, until the method was explained, as being the sort of objects that would probably be turned in a lathe.

But the solution is, that this copying principle admits of endless application, and it will be seen that if we place two lathes side by side, and cause the actual cutter in the one to copy exactly the form which an imaginary cutter is tracing out upon a model in the other, we shall reproduce upon a piece of wood placed in the second lathe the precise pattern which exists as the copy.

In the drawing the mandrels of the two lathes are shown at F and G, the dark oval at F representing a section of the spoke of a wheel, and being, in fact, an exact copy in iron of the thing to be manufactured; the spoke F is attached to the wheel A, while B is an *intermediate* wheel or driver, and C is another wheel of the same size as A.

Fig. 139.

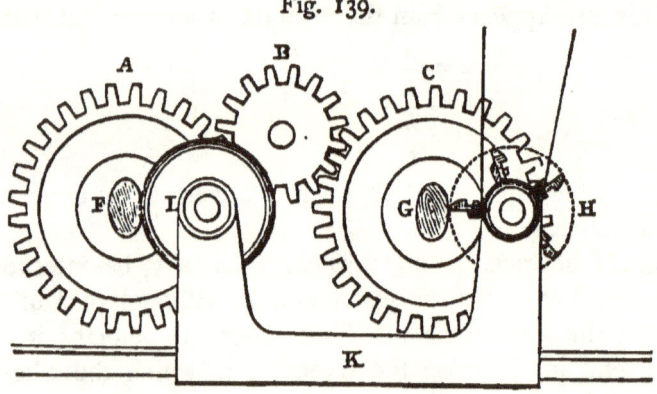

The unfinished spoke is placed parallel to the copy, along the axis of the wheel C, and the function of the intermediate wheel, or driver B, is to cause the material to revolve in the same direction and at the same rate as the pattern.

A sliding frame, K, carries a tracing wheel. I, with a

blunt edge, which is kept pressed against the pattern by a weight or spring, and also contains the cutters, H, which are driven at a speed of about 2,000 revolutions per minute by an independent strap.

The circle described by the extremities of the cutters is precisely the same size as the circle of the tracer, and it follows that the exact form which the tracer feels, as it were, upon the pattern, will be reproduced by the whirling of the cutters against the material, G, and that the spoke may be completed by giving a slow motion to the combination in a direction parallel to the axis of the pattern.

Sometimes the tracer and cutters are mounted upon a rocking frame, instead of upon a slide rest, but the principle of the machine is not changed thereby.

An intermediate wheel may also be useful when two parallel axes are so close together that there is not space for the ordinary spur wheels.

In such a case the axes A and C may be connected by a third wheel, B, and will of course revolve in the same direction.

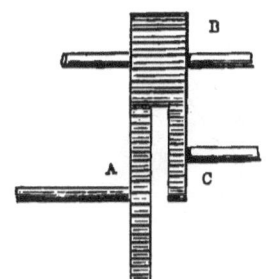

Fig. 140.

The wheel, B, is elongated so as to gear with both A and C, and is called a *Marlborough wheel*.

96. A common eight-day clock affords a familiar illustration of the employment of a train of wheels. (Fig. 141.)

We have marked the disposition of the wheelwork in a clock of this character, and the various wheels are named in the sketch.

The great wheel turns round once in 12 hours, and may have 96 teeth; suppose it to engage with a pinion of 8 teeth on the axis or arbor of the centre wheel, this pinion will turn twelve times while the great wheel turns once, and is capable of carrying the minute hand.

Let the pendulum swing 60 times in a minute, or be a

seconds' pendulum, the scape wheel will then have 30 teeth, and will be required to turn once in a minute.

Hence the value of the train from the centre to the scape wheel should be 60; and in constructing the train we observe that if the pinions on the axes of the second and scape wheels have each of them 8 teeth, the centre and second wheels may have 64 and 60 teeth. In such a case we should have

$$e = \frac{64 \times 60}{8 \times 8} = 60.$$

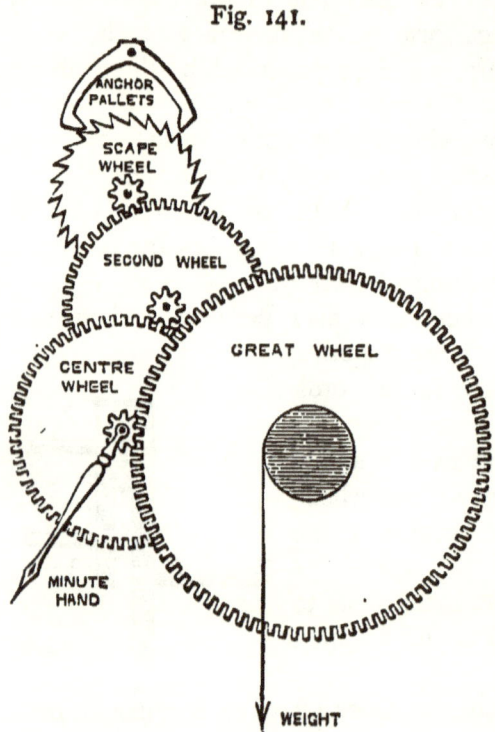

Fig. 141.

In order that the clock may go for 8 days, the great wheel must be capable of turning 16 times before the maintaining power is exhausted.

It is easy to see that if the speed of the scape wheel at one end of the train be increased, and if we are at the same time limited in respect of the number of rotations of the great wheel, it will be convenient to introduce a new axis into the train; and accordingly an additional wheel and pinion is found in the train of a watch, where the balance wheel, which performs the function of a pendulum, makes at least 120 vibrations in a minute.

Another illustration of a train of wheels is found in the method of driving the hour hand of a clock or watch, and in order to understand it we have only to observe that in a clock or watch the minute hand is fastened to the arbor or axis of the centre wheel, and that the hour hand is attached to a pipe which fits upon this axis, and derives its motion from the minute hand.

This appears from the last article, and all we have to do is to connect the pipe and axis by a train of wheels which shall reduce the velocity in the ratio of 1 to 12.

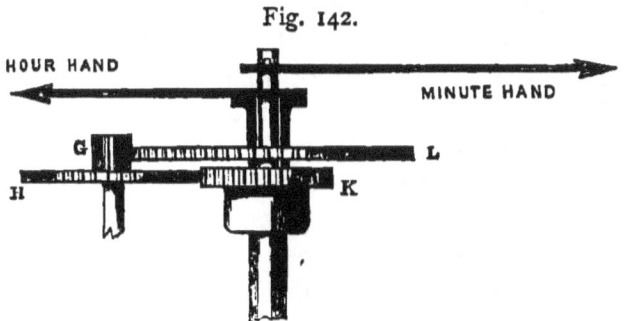

Fig. 142.

The drawing is taken from a small clock, and represents the train of wheels employed. The pinion K, attached to the axis of the minute hand, drives H, whence the motion passes through G to L, and thus to the hour hand, which is fastened to the pipe on which L is fitted, and which corresponds to the mandrel of the lathe. The value of e in the train is given by the equation

$$e = \frac{K \times G}{H \times L} = \frac{28 \times 8}{42 \times 64} = \tfrac{1}{12}.$$

97. It is a maxim among mechanics, that all screws which are required to be perfectly accurate must be cut in a lathe, and there is a geometrical reason for this statement, depending upon the varying inclination of the screw surface at different distances from its axis.

In cutting a screw-thread upon a bolt without using a lathe, we employ pieces of a nut which would exactly fit the screw when finished, in order to carve out the thread.

These pieces, which are called *dies*, are made of soft steel in the first instance, but are afterwards hardened and tempered, and have cutting edges; they are pushed forward by wedges towards the axis of the bolt, during the operation of cutting the thread.

It follows that the angle of a ridge upon the die or cutter begins to trace out the screw-thread upon the bolt. But this angle corresponds to the *inside* line, in the hollow between two ridges, when the screw is completed. We begin, therefore, by tracing out a line, very slightly different in inclination from the line of the thread which we require. The inclination of the thread, when the cutter begins its work, is not theoretically the same as when it leaves off. The difference is scarcely appreciable, or even recognisable, in small screws; but it exists notwithstanding, and we encounter in screw cutting a practical difficulty which has never been absolutely overcome.

We can only avoid this difficulty by having recourse to the lathe.

Here the *copying principle* receives one of its most valuable applications. The maker of a lathe furnishes a screw, shaped with the greatest care and exactness, and places this screw in a line parallel to the bed of the lathe.

The lathe now carries within itself a copy, which can be reproduced or varied at pleasure, for by means of it we can advance the cutter so as to carve out any screw that we may require.

98. The principle of construction of the *screw-cutting lathe* will be apparent from the sketch.

The screw-thread which forms the copy is traced upon the axis C D, and has a definite pitch assigned to it by the maker.

This screw carries a nut, N, and, disregarding the actual construction, we will suppose that the nut,

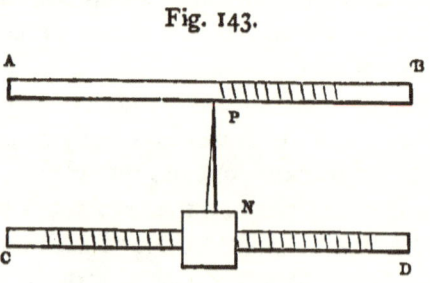

Fig. 143.

N, is furnished with a pointer, P, capable of tracing a screw-thread upon another axis, A B.

Conceive, now, that A B and C D are connected by a train of wheels in such a manner that they can revolve with any required relative velocities.

Upon each revolution of C D the nut advances through a space equal to the pitch of the screw. If A B also revolve at the same rate as C D, and in the same direction, the point P will describe upon A B a screw-thread exactly similar to that upon C D. If A B revolve more or less rapidly than C D, the pitch of the screw upon A B will be less or greater than that upon C D.

Fig. 144.

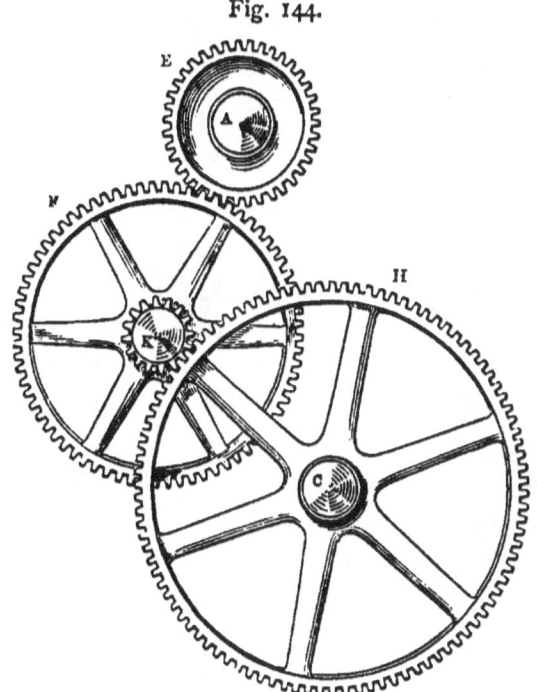

A train which would conveniently connect the axes is shown in Fig. 144. Here C is the axis of the leading screw, and A carries the bar which is to be the subject of the operation; it is, in fact, the mandrel of the lathe.

Let E, F, K, H represent the numbers of the teeth upon the wheels so distinguished, and let e be the value of the train, and suppose A B to make (m) revolutions for (n) revolutions of C D,

we have therefore $\dfrac{\text{pitch of screw on A B}}{\text{pitch of screw on C D}} = \dfrac{n}{m} = e.$

But $e = \dfrac{E \times K}{F \times H}$ ∴ $\dfrac{\text{pitch of screw upon A B}}{\text{pitch of screw upon C D}} = \dfrac{E \times K}{F \times H}.$

A set of *change wheels* is furnished with these lathes, and a table indicates the wheels required for cutting a screw of any given number of threads to the inch. The screw upon C D having two threads in an inch, the numbers of teeth to be assigned to E, F, H, K are given in the table of which a specimen is subjoined.

No. of Threads per Inch.	E	F	K	H
12	90	90	20	120
$12\frac{3}{4}$	60	85	20	90
13	90	90	20	130
$13\frac{1}{2}$	60	90	20	90
$13\frac{3}{4}$	80	100	20	110
14	90	90	20	140

Ex. Let the pitch of the screw upon C D be $\frac{1}{2}$ an inch, and let it be required to cut a screw of $\frac{1}{13}$ inch pitch upon A B, or a screw with 13 threads to the inch.

Here $e = \dfrac{2}{13}$

which is satisfied in the following manner:

$$\dfrac{E \times K}{F \times H} = \dfrac{90 \times 20}{130 \times 90}.$$

The guiding screw being right-handed, the above arrangement is suitable for cutting *right-handed* screws.

To cut a *left-handed* screw it is essential that A B and C D shall revolve in opposite directions.

Now A B revolves with the mandrel of the lathe, and therefore the direction of the rotation of C D must be reversed; this is effected by interposing an idle wheel between H and K, which reverses the motion of the guide screw, C D, and makes the nut travel in the reverse direction.

There is a double slot or groove upon the arm which carries K, in order to allow the adjustment of this idle wheel.

In the case of the micrometer screw with 150 threads to the inch, mentioned in the introductory chapter, a lathe is employed for cutting the thread.

The guiding screw has 50 threads to the inch, and the mandrel of the lathe rotates faster than the guiding screw in the proportion of 3 to 1.

This change of velocity is effected by two wheels having these proportions, and connected by an intermediate wheel, the position of the centre of which can be altered so as to suit principal wheels of different sizes.

The cutter which shapes the thread has a fine pointed edge, and the screw is nearly finished in the lathe, but is finally rendered perfect in form by screwing it through a pair of dies. This latter operation has a tendency to alter the pitch of the screw by permanently stretching the metal of which it is made, and should therefore be resorted to as little as possible.

A screw with 150 threads to the inch and furnished with a graduated head reading off to hundredths of a revolution would measure a linear space of $\dfrac{1}{150 \times 100}$ or $\dfrac{1}{15000}$th of an inch.

This degree of minuteness of measurement, remarkable as it is, has been surpassed, in a manner which has astonished everyone, by the *Measuring Machine* of Sir J. Whitworth.

An inspection of this apparatus would show the value of

the screw-cutting lathe as well as the degree of perfection to which the art of screw cutting has been carried.

Here a screw with 20 threads to the inch has a worm-wheel of 200 teeth fixed upon its axis; this worm-wheel is again driven by a tangent worm or endless screw having a head graduated to 250 divisions.

It follows that the rotation of the tangent screw through one division will advance the bed-screw of 20 threads by a space equal to

$$\frac{1}{250} \times \frac{1}{200} \times \frac{1}{20} \text{ of an inch}$$

or $\frac{1}{1000000}$th of an inch.

Results of this character could be extended as far as we pleased in theory, but not in practice. The accuracy and truth of the pieces upon which we rely are so severely tested that the power of human execution soon fails, and hence we can appreciate the interest which this apparatus has awakened.

In the machine alluded to there are two screws each with 20 threads to the inch, capable of being moved to and fro in the same straight line, the measurement being made in the space between their two ends. Of these screws one has the fine motion due to the tangent screw and the other simply turns in its screwed collar.

The principle is the same as that of measuring by ordinary callipers for end measurement, and if we conceive that a standard piece, called a *gravity piece*, being in fact a small bar of steel with its ends formed into parallel plane surfaces is prepared, an observation would be made by bringing two bars with like plane surfaces into contact with the gravity piece at its two ends so as just to hold it suspended.

Any other piece, of equal length with the standard piece, would be sustained in the same manner with the same screw reading, and could be measured exactly.

Screw-cutting Lathe.

In this way the expansion due to heat of an inch bar touched for an instant with the finger has been detected.

Also the movement of ·000001 inch has been indicated by the gravity piece becoming suspended instead of falling, and the piece has fallen again on reversing the tangent screw through two graduations representing ·000002, showing the almost infinitesimal amount of play in the bearings of the screws.

The contrivance sketched in Fig. 145 is found in every large lathe, and is useful in other machinery where it is required to obtain increased power or a diminished speed. It enables the mechanic to change the velocity of the mandrel of the lathe, and gives another simple example of the use of wheels in trains.

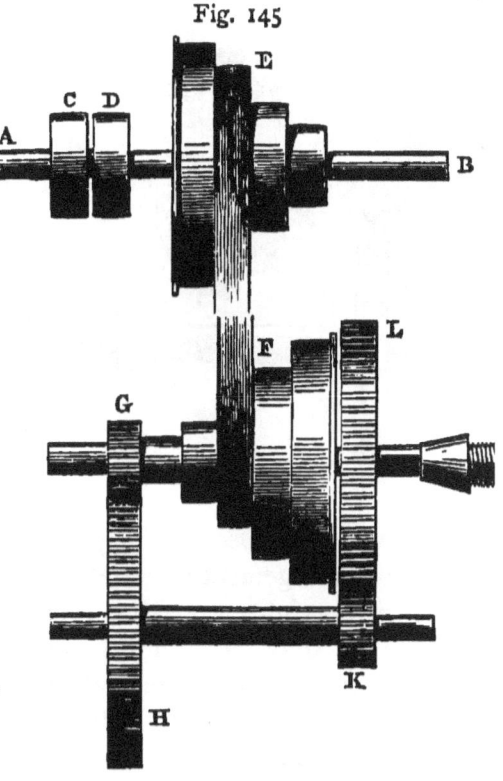

Fig. 145

A B is a shaft overhead, provided with a cone pulley, E, and with fast and loose pulleys at C, D, to receive the power from the engine: a cone pulley F, similar to E, is fitted on the spindle of the lathe, and rides loose upon it; to this cone is attached a pinion G, which drives a wheel H, and so the motion is communicated by the pinion K to the wheel L, which is fastened to the mandrel of the lathe, and turns with it. The result is that the wheel L revolves much more slowly than the cone pulley F, and that the speed of the mandrel is reduced by the multiplier $\frac{G \times K}{H \times L}$, where G, K, H, L represent the numbers of teeth upon these wheels respectively.

Where the lathe is worked at ordinary speeds, the wheels H and K are pushed out of gear by sliding the piece H K in the direction of its axis, as shown in Fig. 146, and the cone pulley, F, is fastened to L by a pin.

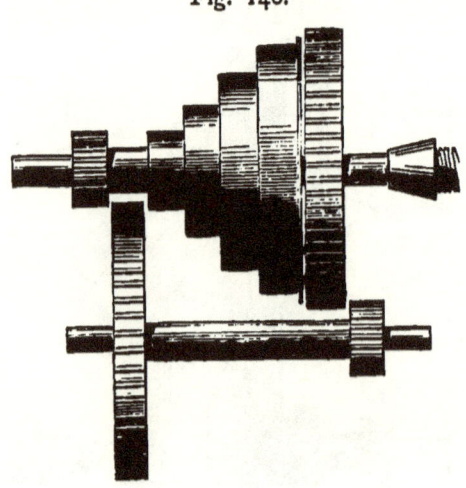

Fig. 146.

This pin must of course be removed as soon as the slow motion comes into work.

As this movement is very similar to the gearing in a crane, we will examine the application of these trains in raising heavy weights, and see how they may be applied so as to reduce velocity, and thereby to increase the amount of force which is called into play.

99. The arrangement of wheelwork in a crane for raising the heaviest weights would be something of the character shown in the drawing, with this difference, that the wheels would be broader and more massive as we approached the axis on which the weight directly acts.

We take a case in which four men, each exerting a force of 15lbs., could raise a weight of somewhat more than 4 tons.

As we are only examining the theoretical power of the combination, we will neglect the loss of power by friction.

Cranes.   143

The men act upon the winch-handles, and the lengths of the arms of these handles are shown as being equal to the diameter of the drum on which the rope or chain is coiled. This gives a leverage of 2 to 1.

We next observe that a large and small wheel are placed upon each axis, and calling e the value of the train, we have the relation,

$$e = \frac{20}{100} \times \frac{40}{120} \times \frac{20}{100}$$

$$= \frac{1}{75}.$$

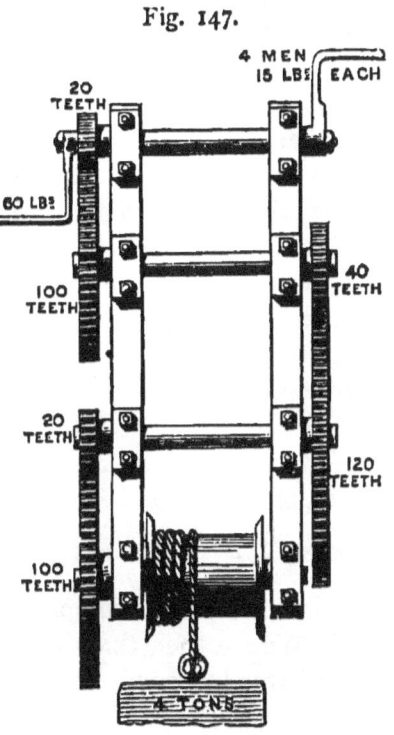

Fig. 147.

Hence the wheelwork multiplies the power 75 times, while the proportion between the length of the winch-handles and the radius of the drum multiplies the power by 2, and thus we reduce the velocity of the weight which is being lifted 150 times as compared with the rate at which the ends of the handles will move; that is to say, the power exerted upon the weight is 150 × 60 lbs., or 9,000 lbs., which is larger than 4 × 2240 lbs., or larger than 4 tons expressed in pounds.

100. We pass on now to an enquiry into the construction of a train of wheels for any given purpose, and here it is necessary to point out that mechanicians are tied down by practical considerations, whereby it often happens that an arrangement which is quite simple and feasible in theory, would nevertheless prove utterly absurd if any attempt were made to carry it out in practice.

One very simple example will explain what we mean.

Suppose it to be required to communicate motion from one axis, A, to another, C, and that C is to make 60 revolutions while A makes 1 revolution, as in the clock train. If A be made to drive C directly, it is clear that the number of teeth upon A must be 60 times as great as the number upon C, so that if C have 8 teeth, A must have 480 teeth.

This would involve the use of two wheels side by side, one of which was 60 times as large as the other, to say nothing of the practical difficulty of dividing the larger wheel so as to form the teeth, and accordingly no such combination is to be found in any clock train.

But the insertion of an intermediate axis relieves us at once from the difficulty.

Place such an axis, which we may call B, between A and C, and fasten upon it two wheels of 8 and 60 teeth respectively, give 64 teeth to A and 8 teeth to C and the value of the train becomes $\frac{64 \times 60}{8 \times 8}$ or 60, the necessary result being obtained with perfect ease and complete simplicity of construction.

101. We see, then, that this problem of connecting two axes by a suitable intermediate train of wheels is an arithmetical problem which may, of course, in some cases prove extremely troublesome, and may demand a considerable amount of arithmetical ingenuity.

The value of e being assigned as a fraction, the only thing to be done is to resolve the numerator and denominator into their prime factors, and then to compose the best train which may suggest itself.

Thus, let it be required to connect two axes so that one shall revolve n times while the other revolves once.

Assume some value for n, say 720,

$$\text{Then } e = n = 10 \times 9 \times 8$$
$$= \frac{80 \times 72 \times 64}{8 \times 8 \times 8}$$

which gives a probable solution for the train.

If any of the factors appear unmanageably large, we may approximate to the value of e by continued fractions, and seek other factors which present less difficulty. If the value of e be an integer, we have seen that it must still be split up into factors, and must be further multiplied and divided by the numbers of teeth in each pinion.

Thus, suppose the two axes are to be connected whereof one revolves in 24 hours, and the other in 365 days 5 hours 48 minutes 48 seconds, as in Mr. Pearson's orrery,

Since 24 hours = 86400 seconds,
and 365 days 5 hrs. 48 min. 48 sec. = 31556928 seconds,

$$\therefore e = \frac{31556928}{86400}$$

$$= \frac{164359}{450}$$

$$= \frac{269 \times 47 \times 13}{10 \times 9 \times 5}.$$

Here 269 is an inconveniently large number, and 5 is certainly too small.

The wheel of 269 teeth cannot be got rid of without altering the entire ratio, but the pinions of 9 and 5 teeth may be changed into others of 18 and 10 teeth.

Thus we have $e = \dfrac{269 \times 26 \times 94}{10 \times 10 \times 18}$.

We might have approximated to e by an algebraical process and have derived the fraction

$$\frac{94963}{260}$$

as representing $\dfrac{31556928}{86400}$ very closely.

But $\dfrac{94963}{260} = \dfrac{11 \times 89 \times 97}{4 \times 5 \times 13}$

$$= \frac{44 \times 89 \times 97}{8 \times 10 \times 13}$$

which avoids the higher number 269, and corresponds to a period of 365 days 5 hours 48 min. 55·38 sec.

L

Thus we have a train in which the numbers are suitable.

Every possible arithmetical artifice is resorted to in cases of this kind, and the ratio $\frac{269}{1}$ has been dealt with after the following manner, in virtue of the discovery that $269001 = 9 \times 9 \times 9 \times 9 \times 41$, and it is not very difficult to get upon the necessary track, for we see at once that 269001 is divisible by 9 because the sum of its digits is so divisible, and again $\frac{269001}{9} = 29889$, which is again divisible by 9 for a like reason, and thus we soon arrive at the last quotient after all the successive divisions by 9, viz. 41.

$$\text{Since } \frac{269}{1} = \frac{269000}{10 \times 10 \times 10}$$

$$= \frac{269001}{10 \times 10 \times 10} \text{ very nearly.}$$

The numerator can now be split up into 3 factors which will express the numbers of teeth in the 3 wheels of a train, and we may consider that

$$e = \frac{269}{1} = \frac{269001}{10 \times 10 \times 10} \text{ very nearly}$$

$$= \frac{81 \times 81 \times 41}{10 \times 10 \times 10}.$$

An approximation which would introduce an error of only one revolution in 269000.

102. It is also a matter of enquiry to ascertain the smallest number of axes which may be concerned in the transmission of any required motion, since we do not want to employ more wheels than are necessary.

The smallest number of teeth which are to be allowed upon a pinion must be given, as well as the largest number to be allowed upon any wheel.

Suppose that no pinion is to have less than 6 teeth, and no wheel more than 60, and let us trace the values of e.

With two axes $e = \frac{60}{6} = 10$.

If the numerator be diminished, or the denominator be increased, the resulting value of e is lessened, or, in other words, 10 is the greatest possible value of e when two axes are employed.

With three axes the greatest value of e is $\frac{60 \times 60}{6 \times 6}$ or 100, and with four axes it is 1000, and so on.

Let e have some value between 10 and 100, we observe that three axes will suffice, and that each wheel must have less than 60 teeth in order to reduce e from 100 to 60.

$$\text{Thus } e = \frac{48}{6} \times \frac{45}{6} = 60.$$

Again, let $e = \frac{365}{3} = 121\frac{2}{3}$, and suppose 180 and 12 to be the limiting numbers of teeth upon a wheel and pinion respectively, in the train which is about to be composed, we shall now find that this extension of the limits of the numbers of teeth upon the respective wheels and pinions, will give us the power of arranging the train without increasing the number of axes.

$$\text{Here } \frac{180}{12} = 15$$

$$\text{and } \frac{180}{12} \times \frac{180}{12} = 15 \times 15 = 225.$$

Now $121\frac{2}{3}$ is less than 225, and therefore three axes will suffice, as in the train represented by

$$e = \frac{180}{18} \times \frac{146}{12}.$$

We may work out this arithmetical reasoning by the use of symbols, and then our solution will apply to every case which can occur.

Assume now that p represents the least number of teeth upon a pinion, w the greatest number upon a wheel, and let x represent the number of fractions in e.

If all the fractions making up the value of e were equal to each other and had the greatest admissible value, then e would reach its limiting value, and we should have

$$e = \frac{w}{p} \times \frac{w}{p} \times \frac{w}{p} \ldots \text{to } x \text{ factors} = \left(\frac{w}{p}\right)^x$$

whence $\log e = x(\log w - \log p)$

$$\therefore x = \frac{\log e}{\log w - \log p}.$$

Now x will probably be a fraction, in which case the next integer greater than $x + 1$ will represent the required number of axes.

Ex. Let $e = \frac{365}{3}$, $w = 180$, $p = 12$,

$$\therefore \frac{w}{p} = 15 \quad \therefore x = \frac{\log 365 - \log 3}{\log 15}$$

$$= 1 + \text{a fraction.}$$

Now the integer next greater than $x + 1$ is 3, therefore 3 axes will be required.

We observe that it is not necessary to find the actual value of x, but simply to ascertain the integer next greater than it.

103. It is sometimes a matter of enquiry how often any two given teeth will come into contact as the wheels run upon each other. We will take the case of a wheel of A teeth driving one of B teeth where A is greater than B, and let $\frac{A}{B} = \frac{a}{b}$ when reduced to its lowest terms.

It is evident that the same points of the two pitch circles would be in contact after a revolutions of B or b revolutions of A.

Hence the smaller the numbers which express the velocity ratio of the two axes, the more frequently will the contact of the same pair of teeth recur.

1. Let it be required to bring the same teeth into contact *as often* as possible.

Since this contact occurs after b revolutions of A or a revolutions of B, we shall effect our object by making a and

b as small as possible, that is, by providing that A and B shall have a large common measure.

Ex. Assume that the comparative velocity of the two axes is intended to be nearly as 5 to 2. And first make A $=$ 80, B $=$ 32, in which case we shall have

$$\frac{A}{B} = \frac{80}{32} = \frac{5}{2} \text{ exactly,}$$

or the same pair of teeth will be in contact after five revolutions of B, or two revolutions of A.

2. Let it be required to bring the same teeth into contact *as seldom* as possible.

Now change A to 81, and we shall still have $\frac{A}{B} = \frac{5}{2}$ very nearly, or the angular velocity of A relatively to B will be scarcely distinguishable from what it was originally. But the alteration will effect what we require, for now $\frac{A}{B} = \frac{81}{32}$ which is a fraction in its lowest terms; there will therefore be a contact of the same pair of teeth only after 81 revolutions of B or 32 revolutions of A.

The insertion of a tooth in this manner was an old contrivance of millwrights to prevent the same pair of teeth from meeting too often, and was supposed to ensure greater regularity in the wear of the wheels; the tooth inserted was called a *hunting cog*, because a pair of teeth, after being once in contact, would gradually separate and then approach by one tooth in each revolution, and thus appear to hunt each other as they went round.

The clockmakers, on the contrary, appear to have adopted the opposite principle.

Finally, we would remind the reader that everything which we have said here about wheels in trains is true, whatever be the directions of their axes; we only care to know the relative sizes of the pitch circles and the directions in which they turn; any part of the train may be composed of *bevil wheels* without affecting our results.

CHAPTER V.

AGGREGATE MOTION.

104. We have seen that every case of the curvilinear motion of a point is of a compound character, resulting from the superposition of two or more rectilinear motions.

It often happens in machinery that some revolving wheel or moving piece becomes the recipient of more than one independent motion, and that such different movements are concentrated upon it at the same instant of time.

The motion is then of a compound or *aggregate* character, and we propose to classify under the head of Aggregate Motion a large variety of useful contrivances.

We commence with two or three simple examples.

The well-known frame called *Lazy Tongs* is a contrivance depending upon aggregate motion.

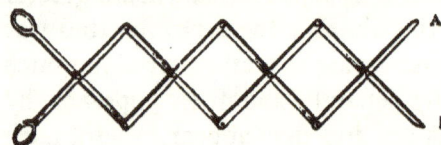

Fig. 148.

The rapid advance of the ends A and B is due to the fact that these points are the recipients of the sum of the resolved parts of the circular motion which takes place at each angle.

Consider the angular joints at the ends of the first pair of bars which carry the handles; these ends of the bars describe circles just as the points of a pair of scissors would do. Either of these motions in a circular arc may be resolved as in Art. 1, and one of the components so obtained will be carried to the end of the combination; the same thing happens at every joint of the series, and thus A and B receive the aggregate of all these separate movements.

Aggregate Motion.

A wheel rolling upon a plane is a case of aggregate motion; the centre of the wheel moves parallel to the plane, the wheel itself revolves about its centre, and these two simple motions give the aggregate result of rolling.

Thus in the case of the driving-wheel of a locomotive, each point on the tyre becomes a fulcrum upon which the rest of the wheel turns, and is for an instant absolutely at rest; the centre of the wheel has the velocity of the train, while a point in the upper edge moves onward with twice that linear velocity. Simple as this matter is, it puzzles some persons when they first think about it.

In some printing machines the table is driven by a crank and connecting rod, and the length of its path may be doubled by applying the principle under discussion.

Here a wheel, Q, is attached to the end of the connecting rod P Q, so that it can turn freely on its centre, Q.

Let the wheel revolve between the two racks, A and B, whereof A is fixed to the framework of the machine, while B carries the reciprocating table.

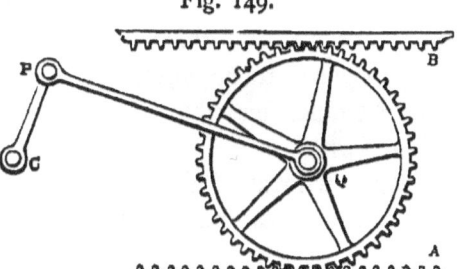

Fig. 149.

The rack B receives the motion of Q in its twofold character, and moves through exactly twice the space that it would describe if connected simply with the point Q.

The size of the wheel makes no difference in the result, for in all cases the velocity of a point in the upper edge will be twice that of the centre.

In the same way, if a beam of timber be moved longitudinally upon friction rollers, the travel of the beam will be twice as great as that of the rollers.

So again, in moving heavy guns, the men employ what is called a *wheel purchase;* that is, they fasten one end of a rope to the spoke of a wheel of the gun-carriage, and make the rope run round the rim: this gives them the leverage of the spokes of the wheel, and the power exerted is exactly one-half of what it would be if the rope were attached directly to the axis of the wheel, in virtue of this principle that the linear velocity of the upper part of the rim is twice that of the centre of the wheel.

105. We may confirm these views of the nature of rolling motion by seeing what would happen if we raised the fulcrum, round which the wheel turns, above the level of the road.

We have now a contrivance by which a carriage may be made to move faster than the horse which draws it, a startling method of stating the fact which has been sometimes adopted. The inventor was a Mr. Saxton, who patented a *Differential Pulley* in the year 1832, with a view of obtaining great speed in railway carriages propelled by a rope; by the use of this invention, the consumption of the rope, proposed to be wound up at a stationary engine house, would be much less than if the carriage were attached in the ordinary way.

Let two wheels of different diameters (say as 6 to 7) be centred on a common axis at c and be fastened together,

Fig. 150.

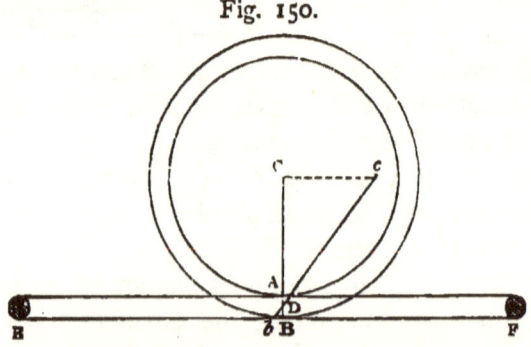

and let an endless rope be wound round the wheels and pass over pulleys at E and F in the manner shown in the diagram, the rope taking a turn round each of the pulleys.

Differential Motions. 153

Conceive now a pull to be exerted on the rope at A in the direction A F, then the tension of the string will cause an equal and opposite pull to be felt at B in the direction B E, and thus the compound pulley has a tendency to turn about D, the middle point of A B.

This tendency in the pulley to turn about the point D causes the linear motion of C to be very much greater than that of any point in the rope; for example, when B moves through a small space, B *b*, the centre C will advance through C *c*, which upon our supposition is thirteen times as great, so that when one yard of rope is wound up, the carriage will have travelled through 13 yards.

The carriage may be at once stopped by disconnecting the pulleys.

106. The *differential screw* is another instance, and is a favourite with writers on mechanics, inasmuch as it gives theoretically a mode of obtaining an enormous pressure by the action of a comparatively small force.

It is constructed on the following principle; two screw threads of different degrees of inclination are formed upon the same spindle A B, the spindle itself passing through two nuts whereof one, E, is part of a solid frame; and the other, D, can slide in a groove along the frame. Let P, Q, repre-

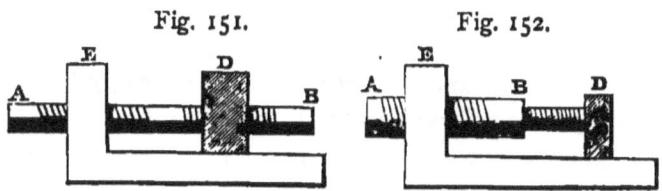

Fig. 151.   Fig. 152.

sent the pitches of the screws at E and D; then upon turning A B once the nut D is carried forward through a space P, and is brought back again through a space Q; it therefore advances through the difference of these intervals.

There is a form of the differential screw described in the fifteenth volume of the 'Philosophical Transactions,' which is known as *Hunter's Screw*. Here one screw is a hollow tube acting as a nut for the second screw in the manner shown in Fig. 152; the smaller screw is attached to a piece D sliding in the frame, and is not allowed to rotate; upon turning the screwed pipe A B, the piece D will move through a space equal to the difference of the pitches of the two screw threads.

If one screw thread were right-handed and the other left-handed, the nut would travel through a space, P+Q, upon each revolution.

107. A right and left-handed screw are often seen in combination, for the purpose of bringing two pieces together.

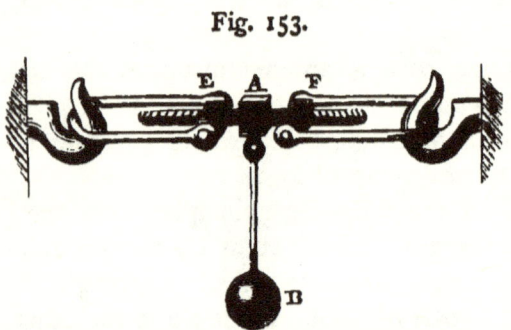

Fig. 153.

There is a very common instance in the coupling which is used to connect two railway carriages. Upon swinging round the arm A B, the screws which are moved by it bring the nuts E and F at the ends of the coupling links closer together, or cause them to separate. This is obviously a most convenient arrangement.

The lever arm and weight at B serve a twofold purpose, they enable the railway servant to screw up the combination easily, so as to put a pressure upon the buffer-springs, and the weight B prevents the screws from shaking loose during the running and vibration of the train.

There is another instance of the use of a right and left-handed screw in combination which is found in the valve-motion of Nasmyth's steam-hammer.

Here a right and left-handed screw are placed side by side and are connected by spur-wheels so that they rotate in opposite directions. Two nuts fastened together engage with the separate screws, and both rise and fall at the same time, being both advanced in the *same* direction by screws which rotate in *opposite* directions.

108. Another contrivance for lifting heavy weights by a small expenditure of power is the *Chinese Windlass*.

Fig. 154.

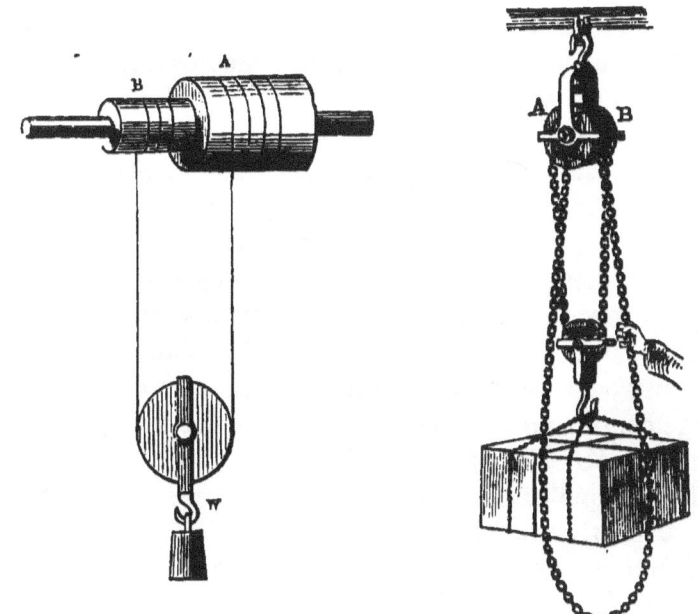

Here a rope is coiled in opposite directions round two axles A and B, of unequal size; the rope is consequently unwound from one axle while it is being wound up by the other, and the weight may rise as slowly as we please.

Let R, r be the radii of the axles, then w moves through $\pi(R-r)$ upon each revolution of the axles.

The practical objection to this windlass consists in the great length of rope required during the operation.

In the ordinary windlass the amount of rope coiled upon the barrel represents the height through which the weight is raised, whereas here we begin by winding as many coils on the smaller barrel as the number of turns which we intend to make with the winch-handle, and then at the close of every turn, a length of rope equal to $2\pi R$ is coiled upon the larger barrel, by which expenditure the weight has only been lifted through $\pi(R-r)$.

Ex. Let $R = 11$, $r = 10$, then the amount of rope wound up in any number of turns bears the same proportion to the space through which the weight is raised that 22 bears to 1.

This is a sufficient commentary on the invention regarded as a practical arrangement.

The object of *Weston's Differential Pulley-block* is to avoid this difficulty about the expenditure of rope. In the Chinese Windlass, one end of the rope is supposed to be fastened to the axle A, and the other end to the axle B; if, however, these two ends were brought together, the supply of rope necessary for B might be drawn from that coiled upon A, and the expenditure would be really $2\pi(R-r)$. There would be many inconveniences attending this arrangement in practice, but it has been put into a working shape by Mr. Weston.

In his pulley-block (Fig. 154) there are two pulleys A and B, nearly equal in size, turning together as one pulley, and forming the upper block; an endless chain supplies the place of the rope, and must of course be prevented from slipping by projections which catch the links of the chain. The power is exerted upon that portion of the chain which leaves the larger pulley, the slack hangs in the manner shown in the sketch, and the chain continues to run round till the weight is raised; the combination is therefore highly effective.

109. The subject of *Epicyclic trains* will now occupy our attention, and we shall discuss some of the most useful applications of that peculiar arrangement of wheelwork which is technically so designated.

An epicyclic train differs from an ordinary train in this particular: the axes of the wheels are not fixed in space,

Epicyclic Trains.

but are attached to a rotating frame or bar, in such a manner that the whole train of wheels can derive motion from the rotation of the bar.

There are certain fundamental forms which consist of trains of two or three wheels; the first wheel of the train is usually concentric with the revolving arm, and the last wheel may be so likewise.

It should, however, be understood that any number of intermediate wheels may exist between the first and last wheels of the train, and that the wheels in the train may derive the whole of their motion from the arm; or they may receive one portion from the arm and the remainder from an independent source.

The elementary form of a train is exhibited in the annexed diagram, and the peculiarities which result from compounding any independent motion with that which arises from the rotation of the arm will demand some careful and attentive study.

Here it will be seen that the wheel B, or the wheels B and C, are attached to a bar which is capable of revolving

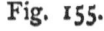

Fig. 155.

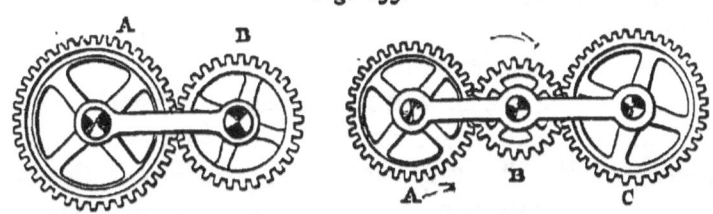

about the centre of the wheel A, the axis of this latter wheel being firmly held in one position.

110. In order to understand movements of this kind let us take a simple case to begin with.

Suppose that there were only two wheels in the train, viz., A and B, and let A be locked so that it cannot rotate; suppose, further, that A has 45 teeth, and that B has 30 teeth, and let us enquire how many rotations B will make while the arm is carried round once.

We might at first imagine that the wheel B would rotate $\frac{45}{30}$ or $\frac{3}{2}$ times by running round upon A; but this is only a part of its movement; the wheel B has also been carried round in a circle about A by reason of its connection with the arm, and having turned upon its axis once more on that account, it has really made $\frac{5}{2}$ turns, instead of $\frac{3}{2}$, during one revolution of the arm.

Another way of considering the subject is the following:

Suppose A to be removed altogether and B to be locked to the arm.

As the arm goes round, a person inspecting the apparatus from a little distance will see the wheel B turning on its axis, and if he watches a mark upon the rim, he can entertain no doubt about this fact.

The truth is that although the wheel B does not move relatively to the arm, it is, nevertheless, the subject of two distinct motions, whereof one consists in a rotation about an axis through its centre, and the other is a motion of *translation*, as it is called, whereby the centre of the wheel describes a circle whose radius is the distance between the centres of A and B.

This is an example of the resolution of a compound movement into its simple elements.

If, again, the wheel B were looked at from the centre about which the arm revolves, no motion of rotation could be recognised. The very same thing happens in the case of the moon. Astronomers tell us that only about one-half of the face of the moon has ever been seen by those upon the earth's surface, and they explain the fact by saying that the moon turns once upon its axis during the period of a single revolution in its orbit round the earth; or in other words, that it moves as if it were fixed to a solid bar stretching from the earth to the moon.

As a further confirmation of this theory, let us consider the case of three wheels A, B and C, whereof A and C are equal. As the arm goes round, C will turn once in the opposite direction to the arm by the rolling of the wheels, and it will turn once in the same direction as the arm by reason of its connection therewith; the aggregate result being that C will be carried round in a circle without rotating at all upon its own axis.

Fig. 156.

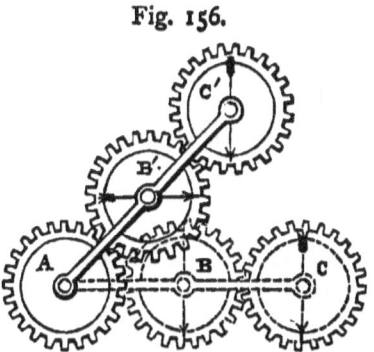

The motions of the wheels B and C in an epicyclic train are shown in the sketch. The arm is supposed to have revolved through an angle of 45°; and it will be seen that B has turned round through a right angle, while C has not rotated at all.

111. We propose now to examine the motion by the aid of analysis.

Remembering that there may be any number of wheels in the train, of which A is the first, and L the last wheel.

Conceive that the arm makes a revolutions ⎫
The first wheel A makes m revolutions ⎬ during the same period
The last wheel L makes n revolutions ⎭ of time,

and let e be the *value* of the train.

Then the first wheel makes $(m-a)$ revolutions relatively to the arm, and the last wheel makes $(n-a)$ revolutions relatively to the same arm, or in other words, L makes $(n-a)$ revolutions for $(m-a)$ revolutions of A.

Recurring to our definition of the *value* of a train, we at once deduce the equality

$$e = \frac{n-a}{m-a}.$$

There are three principal cases to consider:

1. Let A be fixed or $m = 0$,

$$\therefore e = \frac{n-a}{-a} = -\frac{n}{a} + 1$$

$$n = a(1-e) \text{ and } a = \frac{n}{1-e}.$$

2. Let L be fixed or $n = 0$,

$$\therefore e = \frac{-a}{m-a}$$

$$\therefore m = a\left(1 - \frac{1}{e}\right) \text{ and } a = \frac{me}{e-1}.$$

3. Let neither A nor L be fixed, we have now the formula

$$e = \frac{n-a}{m-a}$$

whence $em - ea = n - a$

or $n = me + (1-e)a$.

In applying these formulæ we must consider that e is positive when the train consists of 3, 5, or an odd number of wheels, and negative when there are 2, 4, or an even number in the train.

Ex. 1. Let there be two equal wheels, A and B, in the train, and conceive A to be locked, or let A be a *dead wheel*, as it is termed.

Here $m = 0$, and $e = -1$

$$\therefore n = a(1+1) = 2a,$$

or the wheel B makes two rotations for each revolution of the arm.

Ex. 2. Let there be three wheels A, B and C, whereof A and C are equal, and let A be a dead wheel as before.

Here $m = 0$, $e = 1$,

whence $n = a(1-e) = a(1-1) = 0$,

or C does not turn on its axis at all.

Ex. 3. Take the case first considered, where A is a dead wheel and has 45 teeth, and where B has 30 teeth.

Epicyclic Trains.

Here $e = -\dfrac{3}{2}$

$$\therefore n = a\left(1 + \dfrac{3}{2}\right) = \dfrac{5a}{2}$$

the result arrived at by general reasoning.

112. *Ferguson's Paradox* is obtained by placing three wheels upon the axis which usually carries C, and making these wheels very nearly equal to each other and very nearly equal to A.

Thus let A have 60 teeth, and let the numbers of teeth upon E, F, G, be 61, 60, 59 respectively.

Fig. 157.

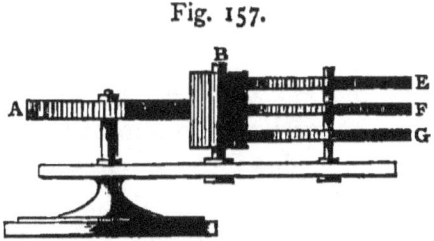

The number of teeth upon B is immaterial, and the wheel A is fixed to the stud upon which it rests, and does not rotate with the arm, so that $m = 0$ throughout the motion.

1. Taking the general formula $e = \dfrac{n-a}{m-a}$, we have, in the case of the wheel E,

$m = 0$, and $e = \dfrac{60}{61}$, which is less than unity.

$$\therefore \dfrac{n-a}{-a} = \dfrac{60}{61}$$

$$\therefore 61\, n - 61\, a = -60\, a$$

whence $n = \dfrac{a}{61}$, and is positive.

2. For the wheel F; $e = \dfrac{60}{60} = 1$,

$$\therefore n - a = -a, \text{ or } n = 0.$$

162   *Elements of Mechanism.*

3. For the wheel G; $e = \dfrac{60}{59}$, which is greater than unity,

$$\therefore \frac{n-a}{-a} = \frac{60}{59}$$

$$\therefore 59n = -60a + 59a,$$

whence $n = -\dfrac{a}{59}$, and is negative.

So that when the arm is made to revolve round the locked or dead wheel A, the wheel E turns slowly in the same direction as the arm, F remains at rest, and G moves slowly in the reverse direction.

This combination has formed a rather popular mechanical puzzle.

113. The *Sun and Planet Wheels* were invented by Watt, and were used to convert the reciprocating motion of the working beam of an engine into the circular motion of the fly-wheel; we have already referred to this invention in Art 4, and have explained the object which it was intended to fulfil. The sketch (Fig. 158) shows the general arrangement of the contrivance; B D is the working beam of the engine, oscillating about the point D, and a wheel A is fixed to the connecting rod B A in such a manner that it cannot rotate, though always in gear with another equal wheel C upon the axis of the fly-wheel. As the beam oscillates, the wheel A will run round C, and is prevented from leaving it by the link A C, which is shown separately at K, or otherwise

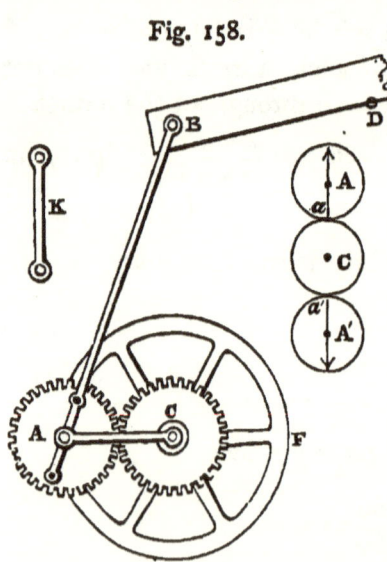

Fig. 158.

by a pin at the back which moves in a groove upon the wheel F. Since it is provided that A shall not turn upon its axis, it follows that C must do so, or the teeth would be torn off, and indeed the rotation of C will be more rapid than we should at first imagine, for it will be found that C makes two complete revolutions upon its axis while A runs round it once.

We may explain the peculiarity as follows: if the discs A and C were fastened together at the point a, and C were to make half a revolution, A would come into the position A' and the direction of the arrow marked upon it would be reversed. But in the actual motion this arrow retains its first direction, and in order to recover it, the disc A' must again rotate through 180°, and must carry C round through another half revolution; so therefore when we recur to the arrangement invented by Watt, C will make a complete revolution while A descends from the highest to the lowest position, or travels half way round it.

If we were to apply our formula $\left(e = \dfrac{n-a}{m-a}\right)$ we should make L the dead wheel, in which case $n = 0$, and $e = -1$,

$$\therefore \quad -1 = \frac{-a}{m-a}$$
$$\therefore \quad m - a = a$$
$$\text{or} \quad m = 2a.$$

which is the result already arrived at.

114. Numerous models intended to illustrate simple astronomical problems connected with the motion of the heavenly bodies are formed upon the principle which is under discussion.

The subjoined arrangement exhibits mechanically the phases of the moon.

Here wheelwork is dispensed with; A and C are simple discs of wood connected by a gutta-percha band, as in the diagram. The band may be open or crossed, according as the discs are required to turn in the same or the opposite direction, and thus we imitate the trains of three or two spur wheels.

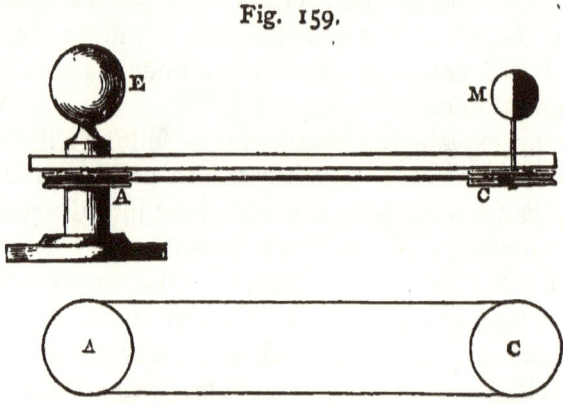

Fig. 159.

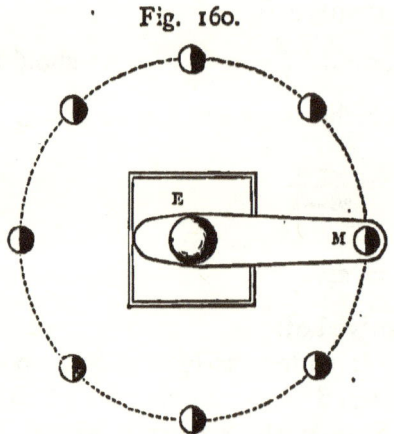

Fig. 160.

E is the earth (Fig. 159); a white ball M, fixed rigidly to the arm, represents the moon; a hemispherical black cap fits upon M, and is connected with the disc C, so as to move with it. If the cap is to represent the dark portion of the moon's surface, it must not rotate as the arm revolves, and this is clearly the case of the wheel F in Ferguson's Paradox; consequently A must be equal to C, and the band which connects A and C must not be crossed.

As the arm revolves, the disc C moves round in a circular path without at all rotating upon its own axis, and the hemispherical cap takes the various positions shown in the sketch, imitating thereby the shadow which would be caused by a luminous body at a great distance to the left of the globe E.

Epicyclic Trains.

115. An epicyclic train may also be formed by the use of three bevil wheels A, B and C, connected as in the figure, and we have now the peculiarity that the wheels A and C turn in opposite directions.

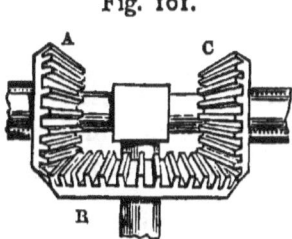

Fig. 161.

The formula already investigated applies equally in this case, and some of the results to be obtained are extremely useful.

Our first example shall be an arrangement whereby the continuation of a piece of shafting may be made to rotate twice as fast as the first portion of it. This forms a simple and easy method of obtaining an increased velocity in a revolving piece, and is used to rotate the coils in some magneto-electric machines.

Thus, let A be a dead wheel, and let B ride loose upon an arm which itself is rigidly attached to the first portion of the shaft, namely, that passing through A, the wheel C being keyed to the other portion which is required to revolve with increased rapidity. As the arm carrying B goes round A, we can easily see that if B were not allowed to rotate at all it would still carry C round once, and that its rotation upon the dead wheel A carries C round a second time, and thus we have an exact reproduction of the motion of the two equal spur wheels, one of which is a dead wheel, and obtain two rotations of C for each revolution of the arm carrying the intermediate wheel B.

We may of course apply the general formula in the case of bevil wheels just as in that of spur wheels, and the expression $e = \dfrac{n-a}{m-a}$ gives the result obtained before.

Thus $m = 0$, $e = -1$, since A and C revolve in opposite directions.

$$\therefore \quad -1 = \frac{n-a}{-a}$$

$$\therefore \quad n = 2a.$$

or C goes round twice for each revolution of the arm which carries B.

116. An illustration may now be taken from the cotton-mills of Lancashire.

During the process of the manufacture of cotton yarn or thread, it is essential to wind the partially twisted fibre upon bobbins, and at the same time this fibre, or *roving*, must not be subjected to any undue strain.

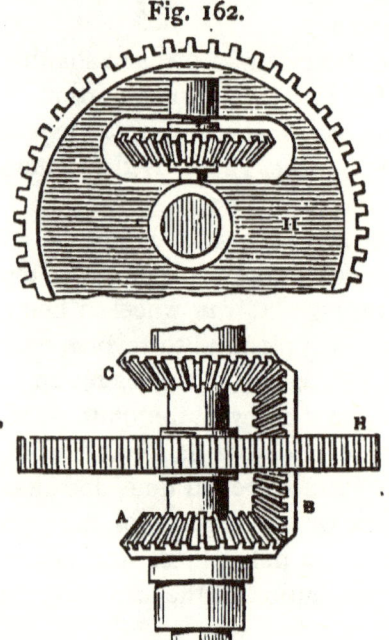

Fig. 162.

The fibre is delivered to the bobbins at a uniform rate, whereas the bobbins get larger as they fill with the material, and hence the winding machinery must be so contrived that the rate of revolution of the bobbin shall slowly decrease upon the completion of each layer of the fibre.

In the year 1826 Mr. H. Houldsworth patented an invention which solves the problem of the *bobbin motion* in the most complete and satisfactory manner.

In the preceding article we have supposed the wheel B to be carried by an arm which is capable of revolving round the axis A C; the better way, however, of suspending B for our purpose is to attach it to the face of a spur wheel, H, as in Fig. 162.

Let this be done, and let A be connected with the driving shaft of the engine, so that its rotation shall necessarily be constant.

If now some independent motion be imparted to the wheel H, the result may be calculated from the formula.

Epicyclic Trains.

Here A, B, C are equal in size, and C rotates in a direction opposite to that of A,

$$\therefore e = -1$$
$$\therefore n - a = a - m$$
$$\therefore n + m = 2a,$$

which gives the analytical relation between the angular velocities of A, C, and H.

If we examine this formula, we shall comprehend that the velocity of C may be reduced by altering the velocity of H.

1. For let $a = m$, or let A and H turn at the same rate, then $n + m = 2a = 2m$

$\therefore n = m$, or C has exactly the same motion as A.

2. Let $a = \dfrac{3m}{4}$, that is, let H make three revolutions while A makes four,

$$\therefore n + m = 2 \times \frac{3m}{4} = \frac{3m}{2}$$

$\therefore n = \dfrac{m}{2}$ or C moves half as fast as A.

3. Let $a = \dfrac{m}{2}$ in which case H makes one revolution for two revolutions of A,

$$\therefore n + m = 2 \times \frac{m}{2} = m$$

$\therefore n = 0$, or C stops altogether.

We have taken extreme cases, from which it appears that when the velocity of the arm is made less than that of A, the velocity of C is reduced in a twofold degree.

Generally, let
$$a = m - \frac{m}{x}$$
$$\therefore n = 2a - m$$
$$= 2m - \frac{2m}{x} - m$$
$$= m - \frac{2m}{x}$$

or the rate of diminution of n is twice that of a.

168   *Elements of Mechanism.*

It now becomes easy to obtain any required reduction in the velocity of C; a reduction in the velocity of H must first be effected by shifting a driving strap along a conical pulley, and the velocity of C will be reduced exactly twice as much as that of H.

Mr. Houldsworth's invention consists, therefore, in imparting to the wheel C two independent motions which travel by different routes, and which, after combination in the manner just investigated, are capable of producing the desired *differential motion*.

117. In order to fix our ideas, let us calculate the motion in the following example:—

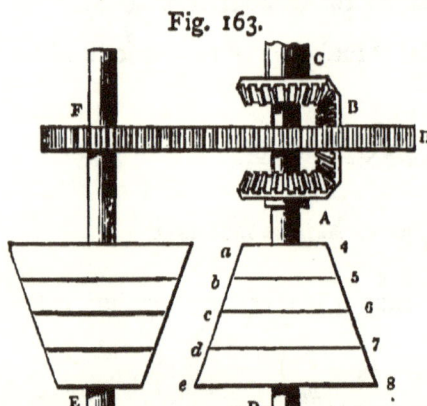

Fig. 163.

Suppose A, B, C to represent three equal wheels, and let A be fixed to a shaft A D, which carries a conical pulley provided with grooves at a, b, c, d, e, where the diameters are 4, 5, 6, 7, 8.

E F is another shaft carrying a second conical pulley which is the counterpart of the first, and terminating in a wheel F whose diameter is half that of H.

A crossed band connects the two cones, and the axis A D is made to revolve with a uniform velocity.

It is required to ascertain the motion of C when the strap is shifted along the conical pulley.

1. Let the strap be placed at a, the velocity of H will be $\frac{1}{4}$ that of A D, and we have $a = \dfrac{m}{4}$,

$$\therefore n = 2 \times \frac{m}{4} - m = -\frac{m}{2}$$

or C moves in the opposite direction to A, and with half its velocity.

2. Let the strap be at b, the velocity of H will be $\frac{5}{14}$ that of A D, (here $e = \frac{5}{7} \times \frac{1}{2} = \frac{5}{14}$ according to Art. 93),

$$\text{therefore } a = \frac{5m}{14},$$

$$\text{and } n = \frac{5m}{7} - m = -\frac{2m}{7}$$

hence c still moves in the opposite direction to A, but less rapidly, in the ratio of 2 to 7.

3. Place the strap at c, when e increases to $\frac{1}{2}$, and a becomes equal to $\frac{m}{2}$,

$$\therefore n = 2 \times \frac{m}{2} - m = 0$$

or c stops altogether, its motion being entirely destroyed.

4. Place the strap at d, and we have $a = \frac{7m}{5} \times \frac{1}{2} = \frac{7m}{10}$

$$\text{whence } n = 2a - m = \frac{7m}{5} - m = \frac{2m}{5}$$

that is, c and A move in the same direction with velocities in the ratio of 2 to 5.

5. Finally adjust the strap at e, and the velocity of H will be the same as that of A D.

Here $a = m$, and $n = 2m - m = m$,

or the motion of c is precisely the same as that of A.

The principle of this invention may now be understood, although it is difficult to appreciate such a movement thoroughly without the assistance of a model.

118. In the manufacture of rope the operation of '*laying*,' or twisting the strands into a perfect rope, has been effected by special machinery.

The Rev. Edmund Cartwright, the inventor of the power-loom, was also the first inventor of a machine for making rope. The general character of the contrivance will be understood from the sketch, which is taken from the specification of the invention.

The machine itself is called a '*Cordelier*,' and consists of a frame placed upon a horizontal shaft P Q, and terminating in a laying-block R, which serves the double purpose of directing the strands to the rollers at K, where they are twisted into rope, and of forming a support or bearing for one end of the shaft.

Three spool frames carry the bobbins, or spools, which contain the supply of strands, and the strands, as they are unwound from the bobbins, pass through delivery rollers at D, E, and F, and thence onward to the laying top.

Fig. 164.

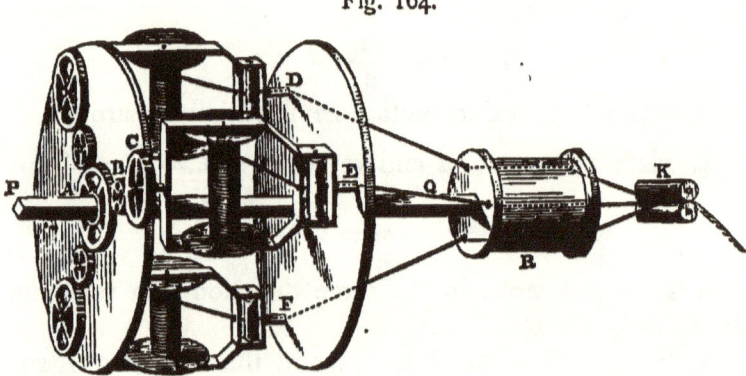

All this is simple enough, and might be the invention of anyone; but there is yet a difficulty to be overcome, which we proceed to explain.

Upon examining a rope it will be found that the twist of the rope is always in the opposite direction to that of the strands, and it follows that if the bobbins were absolutely fixed to the rotating frame the strands themselves would be untwisting during the whole operation. This untwisting is provided against in a rope-walk by the use of two machines, one at each end of the walk; the strands are attached to hooks on one of the machines, and these hooks are made to rotate with a velocity which exactly neutralises

Epicyclic Trains. 171

the twist of the machine which is forming the strands into a finished rope.

In the Cordelier the difficulty is at once removed by the introduction of an epicyclic train. A dead wheel A, so fitted that it remains stationary while the shaft P Q rotates within it, gears with a second wheel B, and this latter with a third wheel C, *equal* to A, whose axis terminates in one of the spool frames. Now we have just proved that in such a train C will run round A without rotating at all upon its own axis, and hence the bobbin may be carried round without in the slightest degree untwisting the strand.

In order to make this matter still more apparent, we refer the student to Fig. 165, which is intended to show three positions of a spool when rotating in a frame without

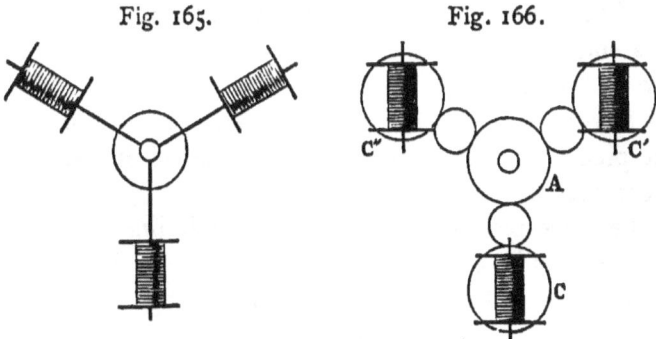

Fig. 165. Fig. 166.

the intervention of an epicyclic train; it is quite evident that the spool has made one rotation round an imaginary axis through its centre while rotating once round the centre of the frame.

In Fig. 166, on the other hand, where an epicyclic train, with C equal to A, is interposed, the bobbin will take the positions C, C', C'', during a revolution, and the rotation just referred to will be exactly neutralised.

119. We have stated that the twist of a rope is always in the opposite direction to that of the strands, and it may be asked, Why is this, and what is the reason that a rope does not untwist itself?

The answer is that any single strand or cord, when twisted up, will always tend to untwist in virtue of the elasticity of its fibres, and that each separate strand in a rope exerts this tendency throughout its whole length, but since the twist of the rope is in the opposite direction, the aggregate of all these comparatively feeble forces is felt as a powerful force restraining the whole rope from becoming untwisted.

It follows, therefore, that by putting a little extra twist upon the strands of a rope in the process of laying, the rope itself will become harder or more tightly twisted.

Captain Huddart incorporated this invention of Cartwright's into some useful machinery for manufacturing rope, and employed the same epicyclic train; but he made the wheel C smaller than A in the proportion of 13 to 14, as in the case of the wheel G in Ferguson's paradox, and the result is that a slight additional twist, or *forehard*, as it is termed, is given to the strands of the rope.

If anyone will try and make a small piece of cord out of three pieces of string he may at once satisfy himself of the correctness of what has been stated.

Take three pieces of string, or fine sash line, thread them through holes in a small plate or disc to keep them separate, and fasten them together at one end, leaving the other ends free.

Upon twisting the knotted end and slowly advancing the disc, a cord will be made which will untwist as soon as it is handled.

Whereas by continually twisting each individual strand, and allowing the knotted end to turn in the opposite direction to that in which the strands are being twisted, a hard piece of cord may be made which will have no tendency whatever to untwist.

120. Another illustration is found in *Equation clocks*. In these nearly obsolete pieces of mechanism the minute hand points to true solar time, and its motion therefore consists of the equable motion of the ordinary minute hand plus or minus the *equation* or difference between true and mean solar time.

In clocks of this class the hand pointing to true solar time is fixed to the bevil wheel.

The wheel A moves as the minute hand of an ordinary clock; the intermediate wheel B is fixed to a swinging arm, E B, as in Art. 115, and the position of C will be in advance of that of A when E B is caused to rotate a little in the same direction, and behind that of A when E B is moved in the opposite direction.

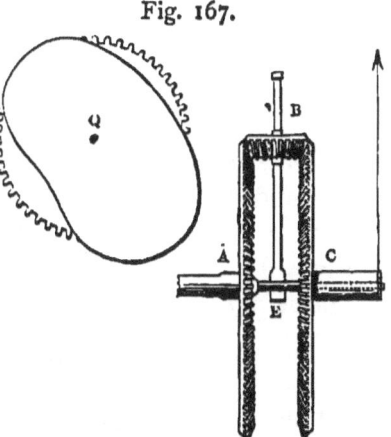

Fig. 167.

Thus as C goes round during each hour of the day, the hand attached to it may be a few minutes before or behind another showing mean time, and deriving its motion at once from A.

The required motion of E B is obtained from a cam plate, Q, curved as in the diagram, and attached to a wheel which revolves once in a year.

121. Epicyclic trains may be employed to produce a very slow motion upon the following principle :—

Let A, B, C, D represent the numbers of teeth in a train of wheels in gear arranged as in the diagram.

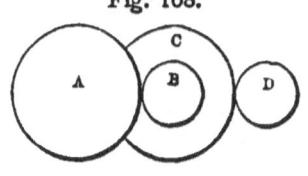

Fig. 168.

If A = D, and B = C, then A and D will rotate with the same velocity in the same direction, but if the equality between (A, D) and (B, C) be slightly disturbed, we shall produce a small change in the *value* of the train.

Suppose, for example, that A is less than D,
or that A = 31, D = 32;

and, again, that B is less than C, or that $B = 125$, $C = 129$: then e, the *value* of the train, will be

$$= \frac{AC}{BD}$$

$$= \frac{31 \times 129}{125 \times 32}$$

$$= \frac{3999}{4000}.$$

Also the more nearly the equality is maintained between (A, D) and (B, C) respectively, the more nearly will the angular velocities of A and D be the same, or the more nearly will e be equal to unity.

Thus if $B = D = 100$, $A = 101$, $C = 99$

we have $e = \dfrac{101 \times 99}{100 \times 100} = \dfrac{9999}{10000}.$

Let us now arrange A, B, C, D in an epicyclic train and carry back the wheel D so that it shall turn upon the same axis as A; the turning of the arm will then set all the wheels in motion except A, which is to be made an immovable or dead wheel, and we shall have D and A moving relatively to each other just as before, that is to say, D will turn very slowly over A at rest.

Fig. 169.

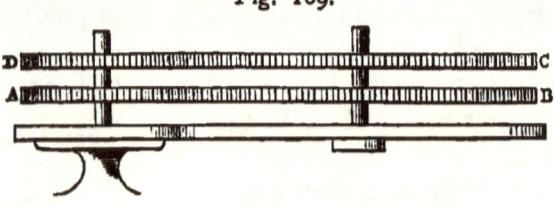

As an easy example take wheels of the following numbers, viz. $A = 60$, $B = 45$, $C = 40$, $D = 65$.

Then $e = \dfrac{AC}{BD} = \dfrac{60 \times 40}{45 \times 65} = \dfrac{32}{39}$

$\therefore 1 - e = 1 - \dfrac{32}{39} = \dfrac{7}{39}.$

If we now rotate the arm and carry round the train, it

Epicyclic Trains. 175

will be found that D makes one revolution when the arm has been carried through a little more than $5\frac{1}{2}$ revolutions, which is also evident from the formula upon observing that $\frac{39}{7} = 5\frac{4}{7}$, which is a little greater than $5\frac{1}{2}$.

So, again, taking the formula $\frac{n}{a} = 1 - e$, and substituting for e the values given previously,

we have in the first example $\quad \frac{n}{a} = \frac{1}{4000}$

and in the second example $\quad \frac{n}{a} = \frac{1}{10000}$.

Hence the arm will make 4000 or 10,000 revolutions respectively while the wheel D turns round once.

122. These examples lead us to compare the movement of any wheel in an epicyclic train with that in another train where the axes are fixed in space, and to regard the subject from a different point of view.

Referring again to the fundamental case, viz. that of three equal wheels A, B, and C, we have seen that if the arm be fixed and A makes one turn, the wheel C will also turn once in the same direction. But if the arm revolve round A fixed, the wheel C will apparently run round just as it did upon the last supposition, and yet at the end of a revolution of the arm it will be found that the wheel C has not turned at all.

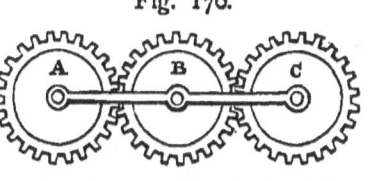

Fig. 170.

The explanation is that the fixed train gives the *absolute* motion of C due to its connection with A, whereas the epicyclic train exhibits the *relative* motion of C with regard to A, which in this case is nothing, because A and C rotate with equal velocities in the same direction.

The same thing is true with respect to any other wheel in the train, such as B; thus when the axes are fixed in space, A and B revolve in opposite directions, and the motion of B *relatively* to A is twice its absolute motion, and thus we account for the fact that in the epicyclic train B will rotate twice while the arm goes round once.

So also in Fig. 169, the fixed train gives the absolute motion of D, viz. $\frac{3999}{4000}$ths of a revolution for each revolution of A, and the epicyclic train exhibits the relative motion of D as compared with that of A, viz. $\frac{1}{4000}$th of the movement of A in the fixed train.

123. The *Parallel Motion* used in steam-engines was the invention of James Watt, and was thus described by himself in the specification of a patent granted in the year 1784 :—

'My second new improvement on the steam-engines consists in methods of directing the piston rods, the pump rods, and other parts of these engines, so as to move in perpendicular or other straight or right lines, without using the great chains and arches commonly fixed to the working beams of the engine for that purpose, and so as to enable the engine to act on the working beams or great levers both by pushing and by drawing, or both, in the ascent and descent of their pistons. I execute this on three principles. . . . . The third principle, on which I derive a perpendicular or right-lined motion from a circular or angular motion, consists in forming certain combinations of levers moving upon centres, wherein the deviations from straight lines of the moving end of some of these levers are compensated by similar deviations, but in opposite directions, of one end of other levers.'

The annexed sketch is copied from the original drawing deposited in the Patent office.

A B is the working beam of the engine, P Q the piston rod or pump rod attached at P to the rod B D, which connects A B and another bar, C D, moveable about a centre at C.

'When the working beam is put in motion the point B describes an arc on the centre A, and the point D describes an arc on the centre C, and the convexities of these arcs, lying in opposite directions, compensate for each other's variation from a straight line, so that the point P, at the top of the piston rod, or pump rod, which lies between these convexities, ascends and descends in a perpendicular or straight line.'

Fig. 171.

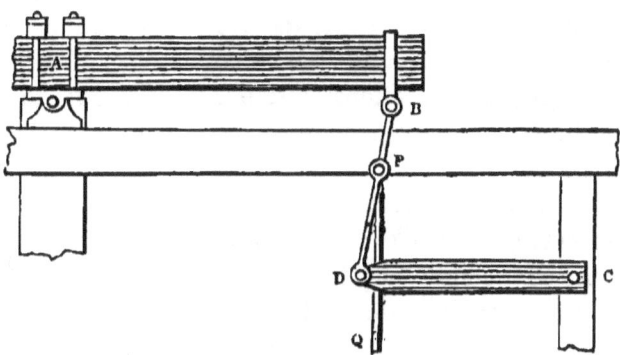

124. This invention being clearly an example of aggregate motion, we proceed now to discuss it in a careful manner, and to examine its peculiar features.

The lines A B and C D in the diagram represent two rods moveable about centres at A and C, and connected by a link, B D. If B D be moved into every position which it can assume, the path of any point P in B D will be a sort of figure of eight, of which two portions, $a\,b$, $c\,d$, are nearly straight lines.

At the beginning of the motion let the rods be so placed that the angles at B and D shall be right angles.

We shall now endeavour to discover that point in B D which most nearly describes a straight line, and in doing so, we first remark that B D begins to shift in the direction of its length,

Fig. 172.

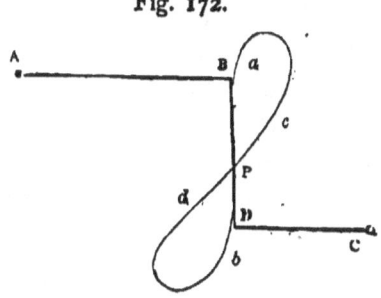

and therefore that the straight line in question must coincide with B D.

The exact position of the so-called *parallel point*, that is, the point P in Watt's diagram, is determined very simply by analysis, and we shall give the investigation immediately. But we can readily predict where it must be found.

As stated by Watt, the points B and D describe circular arcs about the centres A and C, the convexities of these arcs lying in opposite directions, and if A B and C D be equal, the parallel point P must be so placed that its tendency to describe a curve with a convexity approaching to that of the path of B is exactly neutralised by its tendency to describe another curve with a like convexity in the opposite direction due to its connection with the arm C D.

Hence P must lie in the middle of B D, and being solicited by two equal and opposite tendencies, it will follow the intermediate course, which is a straight line. If, however, A B and C D are unequal, the path of the point P will be affected by the increased convexity due to its connection with the shorter arm C D, and in order to escape from this effect it will be necessary to move P away from D, and to bring it nearer to the arm A B, whose extremity traces out a curve of less convexity.

It is pretty clear, since we are dealing with circular arcs, that the point P must now approach B in a proportion identical with that given by comparing A B with C D, or that we must have $\frac{BP}{PD} = \frac{CD}{AB}$, as in Fig. 171.

It is very easy to construct a small model, and to verify in this way the principle of Watt's parallel motion.

125. Refer now to Fig. 173, and suppose the rods to be moved from the position A B D C into another position A $b\,d$ C. Draw $b\,m$, $d\,n$ $\perp^r$ to A B and C D respectively, and let P be the point whose position is to be determined.

Let A B $= r$, b P $= x$, B A $b = \theta$.
C D $= s$, P $d = y$, D C $d = \phi$.

We shall suppose in what follows that the motion of A B and C D is restricted within narrow limits, and shall deal approximately with our equations, by putting

$$\sin \frac{\theta}{2} = \frac{\theta}{2}, \text{ and } \sin \frac{\phi}{2} = \frac{\phi}{2},$$

Fig. 173. Fig. 174.

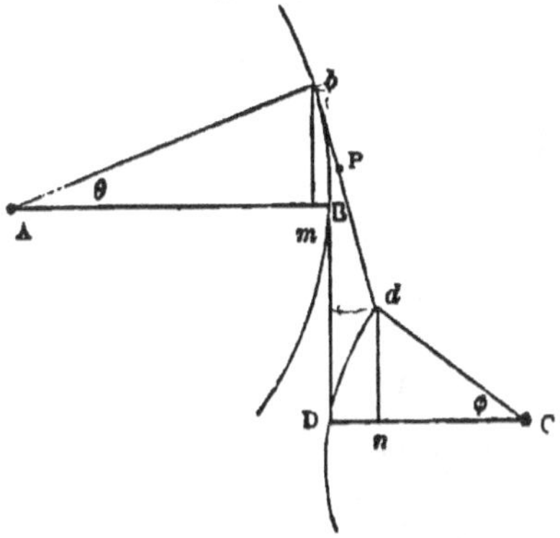

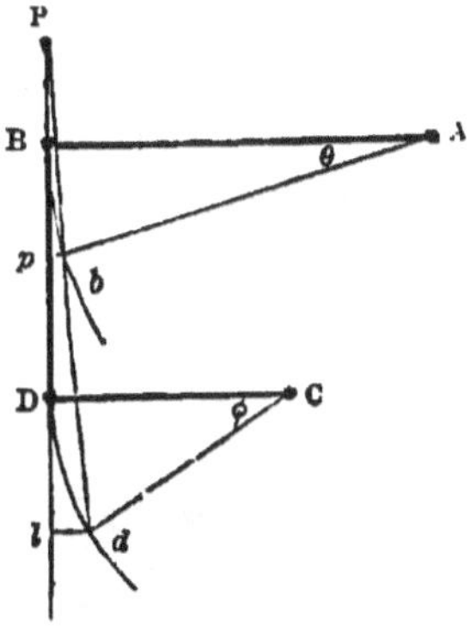

then
$$\frac{x}{y} = \frac{b\,P}{d\,P} = \frac{B\,m}{D\,n}$$
$$= \frac{r(1-\cos\theta)}{s(1-\cos\phi)}$$
$$= \frac{r}{s} \cdot \frac{2\sin^2\frac{\theta}{2}}{2\sin^2\frac{\phi}{2}}$$
$$= \frac{r\,\theta^2}{s\,\phi^2} \text{ nearly.}$$

But the link only turns through a very small angle, which may be considered to be nothing as a first approximation, in which case the vertical motion of B is equal to that of D,

$$\therefore bm = dn \text{ or } r\sin\theta = s\sin\phi$$
whence $r\theta = s\phi$ nearly.

$$\therefore \frac{x}{y} = \frac{r}{s} \times \frac{s^2}{r^2} = \frac{s}{r}$$

or $\dfrac{b\,P}{P\,d} = \dfrac{C\,D}{A\,B}$

i. e. the point P divides B D into two parts which are inversely as the lengths of the nearest radius rods.

In the case considered, which is that which occurs in practice, the parallel point lies in the connecting link, but if the rods be arranged on the same side of the link, as shown in Fig. 174, the required point will lie in B D produced, and on the side of the longer rod.

Suppose now the rods to be moved into the position A $b\,d$ C, and draw $b\,p, d\,l$ perpendicular to B D and B D produced respectively.

$$\text{Then } \frac{b\,P}{d\,P} = \frac{b\,p}{d\,l} = \frac{r\,(1-\cos\theta)}{s\,(1-\cos\phi)}$$
$$= \frac{r\,\theta^2}{s\,\phi^2}.$$

Also $r\,\theta = s\,\phi$, by parity of reasoning,

$$\therefore \frac{b\,P}{d\,P} = \frac{r}{s} \cdot \frac{s^2}{r^2} = \frac{s}{r},$$

that is, $\dfrac{b\,P}{d\,P} = \dfrac{C\,d}{A\,b}$,

and the point P obeys the same general law whether it be found in the link itself or in the prolongation of the line of its direction.

126. We have supposed that $\sin\theta = \theta$ and $\sin\phi = \phi$ in the previous investigation, and have examined the motion of that point in the connecting link which most nearly describes a straight line. We shall now inquire how much P really deviates from the rectilinear path at any given period of its motion.

In practice, the beam of an engine seldom swings through an angle of more than 20° on each side of the horizontal line, and within that limit the error consequent upon our assumption that the sine of an angle is equal to its circular measure would not be considerable; for we find, upon referring to the tables, that the circular measures of angles of 1, 5, 10, 15, 20 degrees, and the natural sines of the same angles are the following:—

Parallel Motion.

Angle.	Circular Meas.	Natural Sine.	Difference.
1°	·0174533	·0174524	·0000009
5°	·0872665	·0871557	·0000108
10°	·1745329	·1736482	·0008847
15°	·2617994	·2588190	·0029804
20°	·3490659	·3420201	·0070458

In an engine of the usual construction A B is equal to C D, and we shall simplify our results by making this supposition.

Let B D move into the position $b\,d$, and turn through an angle, a; it is very apparent that within the limits to which the rods move in practice, a will be much less than θ or ϕ, so that we may regard a as a close approximation to the actual value of $\sin a$ even when we do not adopt the same supposition with regard to θ and ϕ.

The object of the investigation will be to determine ϕ in terms of θ, and we shall see that the deviation sought for depends upon the difference of the cosines of ϕ and θ.

As before, observing that $s = r$, we have

Fig. 175.

$$B\,m = r(1 - \cos \theta)$$
$$m\,b = r \sin \theta$$
$$d\,n = r(1 - \cos \phi)$$
$$D\,n = r \sin \phi.$$

Let $B\,D = l$, then $l + m\,b = l \cos a + D\,n$,
or $l + r \sin \theta = l \cos a + r \sin \phi$
$\therefore r \sin \phi = r \sin \theta + l(1 - \cos a)$ (1).

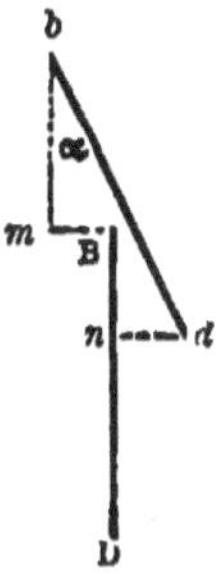

Now a being the angle through which B D is twisted, and being moreover very small, we shall have

$$l\,a = B\,m + d\,n \text{ very nearly}$$
$$= r(1 - \cos \theta) + r(1 - \cos \phi)$$
$$= 2r(1 - \cos \theta), \text{ since } \phi \text{ is nearly equal to } \theta.$$

$$\therefore a = \frac{2r}{l}(1-\cos\theta) \text{ very approximately.}$$

By substituting in equation (1) we can calculate ϕ with considerable accuracy, and then the deviation of P from the vertical line will

$$= \frac{d\,n - \text{B}\,m}{2}$$

$$= \frac{r}{2}(\cos\theta - \cos\phi)$$

and can therefore be ascertained.

Ex. Let $\theta = \dfrac{\pi}{9}$, and assume $r = s = 50$ in., $l = 30$ in.

$$\therefore a = \frac{2r}{l}('0603074)$$

$$= \frac{10}{3}('0603074)$$

$$= '2010247$$

or a represents the angle $11°\ 31'$.

Substituting in equation (1) we have

$$\sin\phi = \sin 20 + \frac{3}{5}(1 - \cos 11°\ 31')$$

$$= '3420201 + \frac{3}{5}('0201333)$$

$$= '3541001$$

$\therefore \phi$ represents an angle of $20°\ 44'$ nearly

Hence the deviation of P from the vertical

$$= \frac{50}{2}(\cos 20° - \cos 20°\ 44')$$

$$= 25\,('0044544)$$

$$= \tfrac{1}{10}\text{th of an inch approximately.}$$

It may be shown that this amount of deviation is again capable of reduction if we cause the centres of motion, A and C, to approach each other by shifting them horizontally through small spaces.

127. The point P, whose motion has been examined, is usually found at the end of the air-pump rod. We have now to obtain a second point, also describing a straight line, and suitable for attachment to the end of the piston rod.

We require, in the first instance, to know when two curves are similar, and in Newton's 'Principia' the test of similarity is stated in the following terms:—

Two curves are said to be similar when there can be drawn in them two distances from two points similarly situated, such, that if any two other distances be drawn equally inclined to the former, the four are proportional.

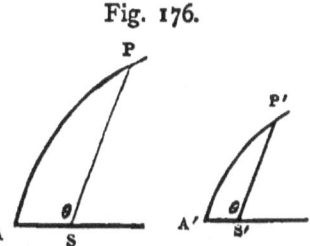

Fig. 176.

Ex. Thus all parabolas are similar curves, and all ellipses with the same eccentricity are similar curves.

Let A, A', be the vertices, S, S', the foci of two parabolas.

Then S A, S' A', are two lines drawn from two points similarly situated, viz. the foci of the curves.

Let S P, S' P' be radii inclined at the same $\angle \theta$ to S A, S' A' respectively.

$$\text{Then } SP = \frac{2\,SA}{1 + \cos\theta}$$

$$S'P' = \frac{2\,S'A'}{1 + \cos\theta}$$

$$\therefore \frac{SP}{S'P'} = \frac{SA}{S'A'},$$

whence the curves are similar, and there is no exception to this rule.

Those who are conversant with the properties of a parabola know very well that it represents, with great exactness, the path of a stone thrown obliquely into the air, and gives the theoretical form of the path of a projectile when unaffected by the resistance of the air.

The similarity of all such curves to one another is by no means evident upon cursory observation, but it is at once established by this simple reasoning.

In the case of ellipses, we proceed in a similar manner, and now s and s' represent the foci of two ellipses of eccentricity e and e' respectively.

$$\text{Here } SP = \frac{SA(1+e)}{1+e\cos\theta}$$

$$\therefore S'P' = \frac{S'A'(1+e')}{1+e'\cos\theta}.$$

Let now $e = e'$, the eccentricities being identical,

$$\text{then } \frac{SP}{S'P'} = \frac{SA}{S'A'},$$

or the curves are similar only under the condition stated.

128. Without any further enquiry into the nature of the curves which satisfy the condition of similarity, we will pass on to examine an extremely useful instrument called a *Pantograph*, which is formed as a jointed parallelogram with two adjacent sides prolonged to convenient lengths, and is used to enlarge or reduce drawings according to scale.

This parallelogram was incorporated by Watt into the invention of the parallel motion, and gave it that completeness which it has at the present time. We have now to show that the pantograph is an apparatus for tracing out similar curves.

In the diagram, let B Q R C represent a parallelogram whose

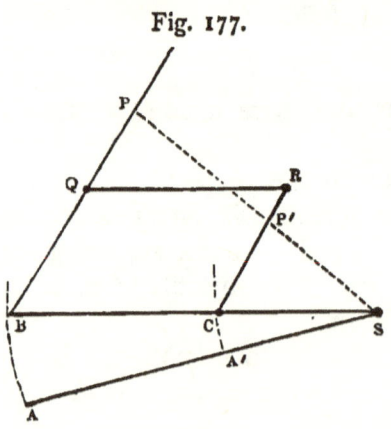

Fig. 177.

sides are jointed at all the angles, and having the two adjacent sides B C, B Q, lengthened as shown. Take a point S, somewhere in B C produced, as a centre of motion, place a pencil at any point P' in the side R C, produce S P' to meet B Q, or its prolongation in P, place another pencil at P, when it will be found that by moving about the frame over a sheet of paper, and at the same time allowing the joints free play, it will be possible to describe any two curves that we please, and these curves will be similar to each other.

Our definition tells us that the two pencils will trace out similar curves if we can show that S P always bears to S P' the same ratio that two other *fixed* lines radiating from S, and to which S P and S P' are equally inclined, also bear to each other.

Conceive, now, that S C B originally occupied the position S A' A, and draw the line S A' A as a fixed line upon the paper, then we shall always have

$$\frac{S P}{S P'} = \frac{S B}{S C} = \frac{S A}{S A'}, \text{ which is a constant ratio,}$$

and the angle A S P must be equal to the angle A' S P', therefore the points P and P' will trace out similar curves so long as S P' P remains a straight line.

This is therefore the only condition which we have to observe in using the instrument.

129. We are now in a position to complete the *Parallel Motion of a Beam Engine*, for if one of the points describe a straight line, the other must do the same.

To the system, A B C D, we superadd the parallelogram B F E D; A B F being the working beam of the engine.

Fig. 178.

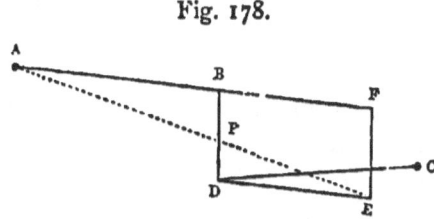

The usual construction is to make the arms A B and C D equal to each other, in which case P, which is the point to which the air-pump rod is fixed, will be in the centre of B D; whereas the second point E, which we are about to find, lies in A P produced, and is the point of attachment of the end of the piston rod.

In order to find the side B F in this parallelogram B F C D, assume that $AB = r$, $CD = s$, $BF = x$.

Then $\dfrac{x}{r} = \dfrac{DP}{PB}$ by similar $\triangle^s$ A B P, D P E.

Also $\dfrac{DP}{PB} = \dfrac{AB}{CD} = \dfrac{r}{s}$, by property of the parallel motion,

$$\therefore \dfrac{x}{r} = \dfrac{r}{s} \text{ or } x = \dfrac{r^2}{s},$$

which equation determines the proportion between B F, A B, and C D, in order that the second point sought for may lie at the vertex, E, of the parallelogram B F E D.

Thus we see that the complete arrangement consists of two distinct portions incorporated together.

1. The combination of A B, C D, and B D, which compels some point P to describe a straight line, the position of this point depending upon the relation between A B and C D.

2. The Pantograph A B F E D, which has some point in F E, here for simplicity selected at E, that must necessarily describe a straight line parallel to the path of P.

And, as we have stated, the points of attachment of the ends of the air-pump and piston rods to the main beam of the engine are thus provided for.

130. In modern engines, the principle of the expansive working of steam is extensively carried out, and there are often two steam cylinders in the place of one, viz. a high-pressure cylinder of small dimensions, and a larger, or low-pressure cylinder by the side of it.

The steam is first admitted into the smaller cylinder, passes from thence into the larger one, and finally escapes into the condenser, as in an ordinary condensing engine.

The use of these double cylinders, with a swinging beam, necessitates a more complex form of parallel motion, but it is quite easy to understand the construction if we remember the principles already investigated.

Parallel Motion.

The diagram shows the arrangement of the Pumping Engine at the Lambeth Water Works.

Fig. 179.

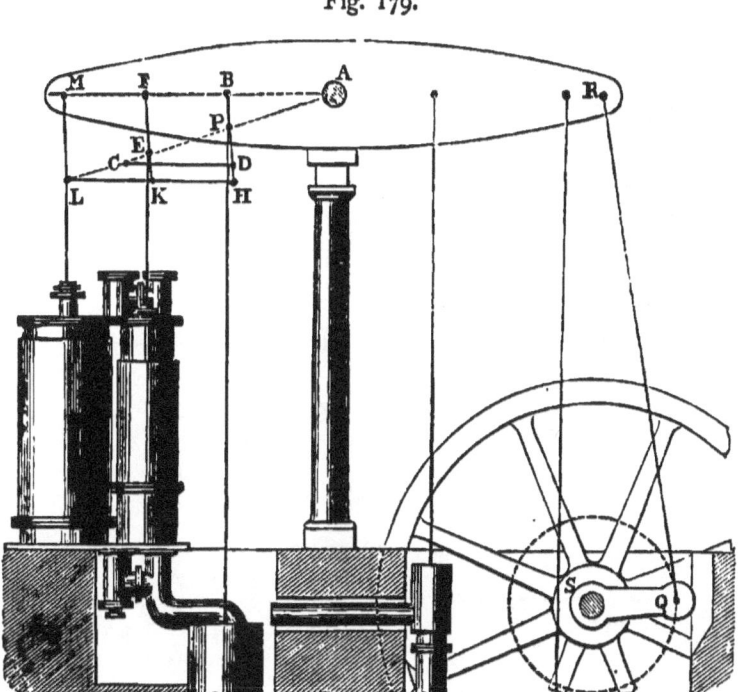

The first part of the parallel motion for connecting the two piston rods with the beam of the engine consists of the portion C D B A, C being a fixed centre of motion, and A being the axis of the beam. In this portion the two arms C D, A B, and the link B D give the parallel point P.

If we now join A P, and produce it to meet the sides of the superadded parallelograms in the points E and L, we shall obtain two other points whose motion is similar to that of P. To these points the ends of the piston rods must be attached, and the arrangement is complete. The addition, therefore, of an intermediate bar, F E K, parallel to the side M L of the ordinary parallelogram, gives us what we require.

131. Where a beam engine is used in a steam vessel the

beam must be kept as low down as possible, and the motion is altered as in the figure, but it is precisely the same in principle.

Beginning with A B D C, a system of two arms and a connecting link, we obtain the parallel point P; we then construct the pantograph C D B F E so as to arrive at the point P', whose path is similar to that of P.

Here C E represents the beam of the engine, and P' is the point to which the end of the piston rod is attached. Draw G H parallel to F E, and let $\begin{matrix} AB = r \\ CD = s \end{matrix} \begin{Bmatrix} DE = c \\ BD = l \end{Bmatrix}$ F P' $= x$.

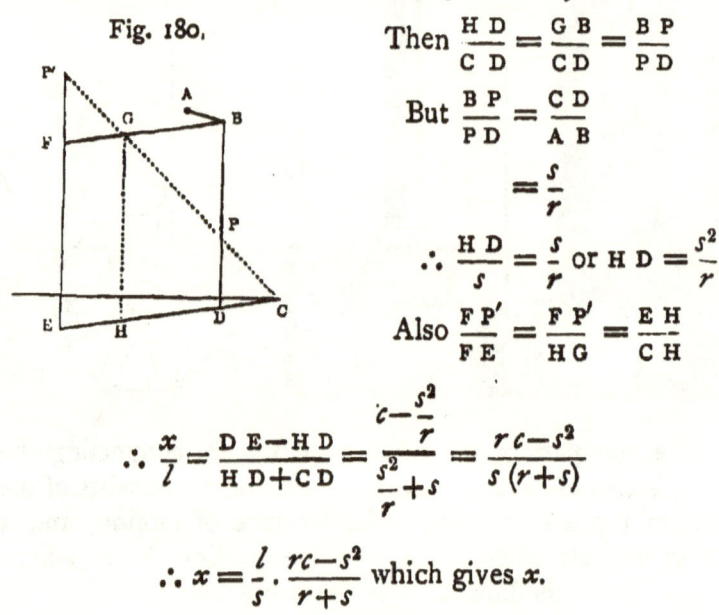

Fig. 180.

Then $\dfrac{HD}{CD} = \dfrac{GB}{CD} = \dfrac{BP}{PD}$

But $\dfrac{BP}{PD} = \dfrac{CD}{AB}$

$= \dfrac{s}{r}$

$\therefore \dfrac{HD}{s} = \dfrac{s}{r}$ or $HD = \dfrac{s^2}{r}$

Also $\dfrac{FP'}{FE} = \dfrac{FP'}{HG} = \dfrac{EH}{CH}$

$\therefore \dfrac{x}{l} = \dfrac{DE - HD}{HD + CD} = \dfrac{c - \dfrac{s^2}{r}}{\dfrac{s^2}{r} + s} = \dfrac{rc - s^2}{s(r+s)}$

$\therefore x = \dfrac{l}{s} \cdot \dfrac{rc - s^2}{r+s}$ which gives x.

132. Another form of parallel motion was devised for marine engines before the principle of direct action was so generally adopted. It was fitted to the engines of the 'Gorgon' by Mr. Seaward, and has since been applied in a modified form to small stationary engines; but except so far as the latter application is concerned, it has not been regarded with particular favour.

It is, however, remarkable as illustrating a mechanical

principle for reducing the friction upon an axis, by causing the driving pressure and the resistance to be overcome to act upon the same side of the centre of motion; for here the connecting and piston rods are both attached to the rocking beam upon the same side of its axis. In this respect it has an advantage, for in ordinary beam engines the pressure upon the fulcrum of the beam is the sum of the power and the resistance, whereas here it is the difference of these forces, and the friction is proportionally diminished.

It is a modification of a simple geometrical fact.

Let the rod A B be bisected in C, and jointed at that point to another rod, C D, which is equal in length to A C. Suppose the point D to be fixed as a centre of motion, and the end A to be constrained to move to and fro in the line A D, then B will move up and down in a straight line pointing also to D.

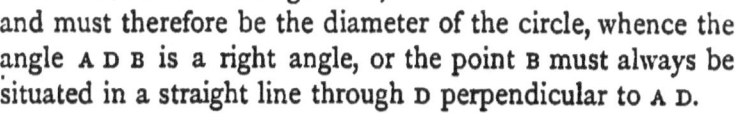

Fig. 181.

Since C A = C D = C B, the point C will be the centre of a circle passing through A, D, and B.

Also A C B is a straight line, and must therefore be the diameter of the circle, whence the angle A D B is a right angle, or the point B must always be situated in a straight line through D perpendicular to A D.

That is, the path of B is a fixed straight line pointing to D.

In the parallel motion of the Gorgon engines the piece corresponding to A B has the end A swinging at the extremity of another bar, so as to describe a small arc very nearly coinciding with a straight line.

Referring to Fig. 182, we have H S P corresponding to B C A, the point P moving very approximately in a horizontal line by reason of its connection with P Q which has a centre of motion at Q, and the point H being that which most nearly describes a straight line. The system of rods is then T S, H S P, P Q, the points T and Q being centres of motion.

Draw S R and H K $\perp^r$ T P, and S V $\perp^r$ H K,

and let $\begin{matrix} TS = a \\ SH = b \end{matrix}\Big\}$ $SP = c,$ $\begin{matrix} STR = \theta \\ SPR = \phi \end{matrix}\Big\}$

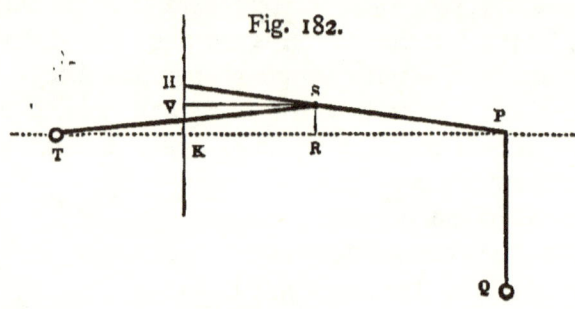

Fig. 182.

Then $TR = a \cos \theta = a \left(1 - 2 \sin^2 \frac{\theta}{2}\right)$

$= a \left(1 - \frac{\theta^2}{2}\right)$ nearly,

$SV = b \cos \phi = b \left(1 - \frac{\phi^2}{2}\right)$ nearly,

$\therefore TK = a - b - \frac{a\theta^2}{2} + \frac{b\phi^2}{2}$

But the point H describes the straight line H K,

$\therefore TK = a - b$

whence we have $\frac{b\phi^2}{2} - \frac{a\theta^2}{2} = 0$

or $a\theta^2 = b\phi^2$

But $\frac{a}{c} = \frac{\sin \phi}{\sin \theta} = \frac{\phi}{\theta}$ nearly,

$\therefore \frac{a}{b} = \frac{a^2}{c^2}$

$\therefore c^2 = ab,$

or T S is a mean proportional between S P and S H, a condition to be fulfilled by the rods giving the parallel motion.

133. A *Parallel Motion* may also be useful in machinery. In the old process of multiplying engraved steel plates at the Bank of England, which was practised before the art

Fig. 183.

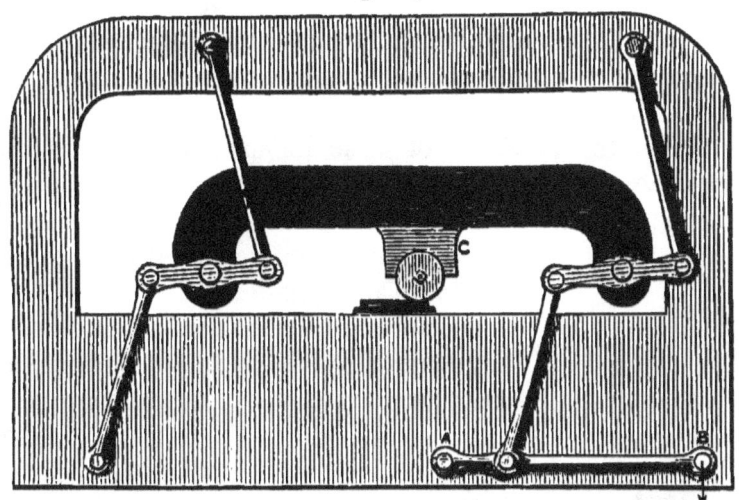

of electrotyping was understood, it was necessary to roll a hardened steel roller upon a flat plate of soft steel with a very heavy pressure, and so to engrave the plate. The difficulty of maintaining this pressure during the motion of the roller upon the surface was overcome by the aid of the parallel motion shown in the drawing.

The system of jointed bars allowed the heavy frame C to traverse laterally, while the necessary pressure was obtained by a pull upon the end B of the lever A B, which lever was moveable round A as a centre of motion, and was further connected at B with some source of power.

134. *Watt's Indicator* is an instrument used to ascertain the actual horse-power of a working steam-engine. The principle upon which it is constructed is the following :—

A pencil oscillates through the space of a few inches in a horizontal line, with a velocity which always bears a fixed ratio to that of the piston, whereby its motion is an exact counterpart upon a very reduced scale of the actual motion of the piston in the steam cylinder; and at the same time it is the subject of a second movement in a vertical line, which is caused by the pressure of the steam or uncondensed vapour in the cylinder, and occurs whenever the pressure of the steam or vapour upon one and the same side of the piston of the engine becomes greater or less than that of the atmosphere.

Under the influence of these independent motions the aggregate path of the pencil will be a curve which is capable of interpretation, and which affords a wonderful insight into actions which are taking place in the interior of the cylinder.

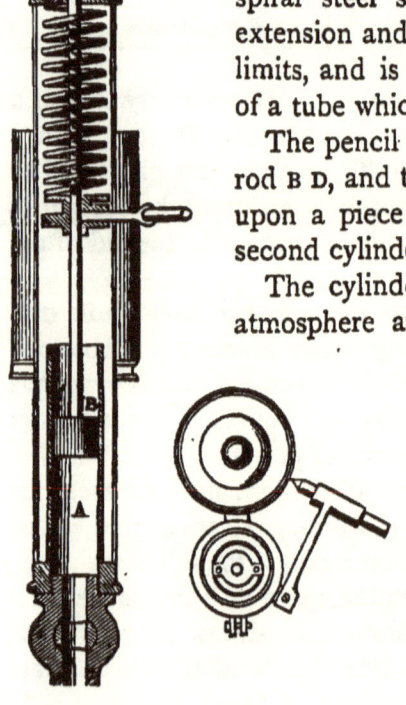

Fig. 184.

The indicator consists of a small cylinder, A, fitted with a steam-tight piston, B; the piston rod, B D, is attached to a spiral steel spring, which is capable of extension and compression within definite limits, and is enclosed in the upper part of a tube which carries the cylinder A.

The pencil is attached to a point in the rod B D, and traces the indicator diagram upon a piece of paper wrapped round a second cylinder by the side of the first.

The cylinder, A, is freely open to the atmosphere at the top, and a stopcock admits the steam when required. The indicator is usually fixed upon the cover at one end of the steam cylinder of the engine. When the stopcock is opened and the lower side of B is in free communication with the interior of the cylinder, the pressure of the steam will be usually greater or less than that of the atmosphere; if it be greater, B will rise against the pressure of the spring, and if it be less, the pressure of the atmosphere upon the upper surface of B will overcome the resistance of the spring and cause the pencil to descend.

At the same time, the cylinder which carries the paper is made to turn with a motion derived at once from that of the piston in the engine, but much less in degree, and thus a curve is traced out somewhat of the character represented below.

Fig. 185.

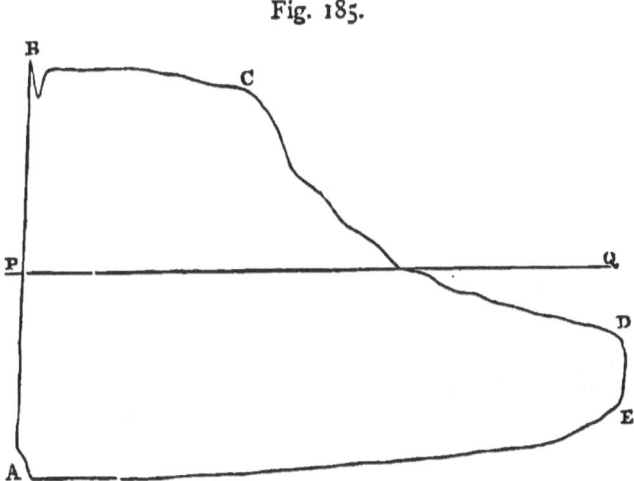

Here P Q is the atmospheric line, and is the path of the pencil when the pressure of the steam is equal to that of the atmosphere, or when the spiral spring is neither extended nor compressed.

As the steam enters the cylinder, the piston may be supposed to be descending, and the pencil to be describing the upper portion of the curve; when the piston returns, the pencil moves to the left through D E A, and thus the diagram is traced out. We may examine this matter with more particularity as follows : the steam is admitted just before the piston reaches the top of its stroke by giving a small amount of lead to the slide valve, and the pencil rises at once with a rapid motion from A to B; the full pressure of the steam is then maintained while the pencil, recording a portion of the travel of the piston, moves from B to C; at C the steam is cut off, and the pencil falls gradually as the steam expands with a diminishing pressure; at D the steam pours into the condenser, and the fall becomes sudden; from E to A the cylinder is in full communication with the condenser, and the pencil describes a line somewhat inclined to the line P Q, the position and form of which depends upon the perfection of the vacuum in the condenser.

The strength of the spiral spring being ascertained, the curve tells us exactly the number of pounds by which the pressure of the steam urges the piston onward during every inch of its path in one direction, and the amount of resistance which the uncondensed vapour or gases existing in the condenser oppose to its passage in the other direction; the area of the curve, therefore, affords an estimate of the work done in the engine during one complete stroke, and is a graphic representation of the same; the engineer estimates this area by simple measurement in the most direct manner which occurs to him, and the actual indicated horse-power is obtained by multiplying the work done in one stroke by the number of strokes made in a minute, and then dividing by 33,000, the number of pounds that a horse can raise through one foot in one minute.

135. The object of the indicator being to ascertain the exact pressure of the steam or vapour in the cylinder at each point of the stroke of the piston, it has been found to be a great advantage to diminish as much as possible the play of the spring which controls the pencil. In this way the vibration and irregularity of motion of the pencil is greatly reduced. But the play given to the spring determines the height of the diagram, and we do not wish to reduce this, but rather the contrary: it is not easy to reconcile these contradictory requirements, but nevertheless a form of indicator has been invented by Mr. Richards which solves the difficulty and has become most deservedly popular.

It is an ingenious application of the combination of two bars and a link forming a parallel motion, and will be understood at once from the drawing, which is taken from a small model representing very closely the essential parts of an actual instrument.

Fig. 186.

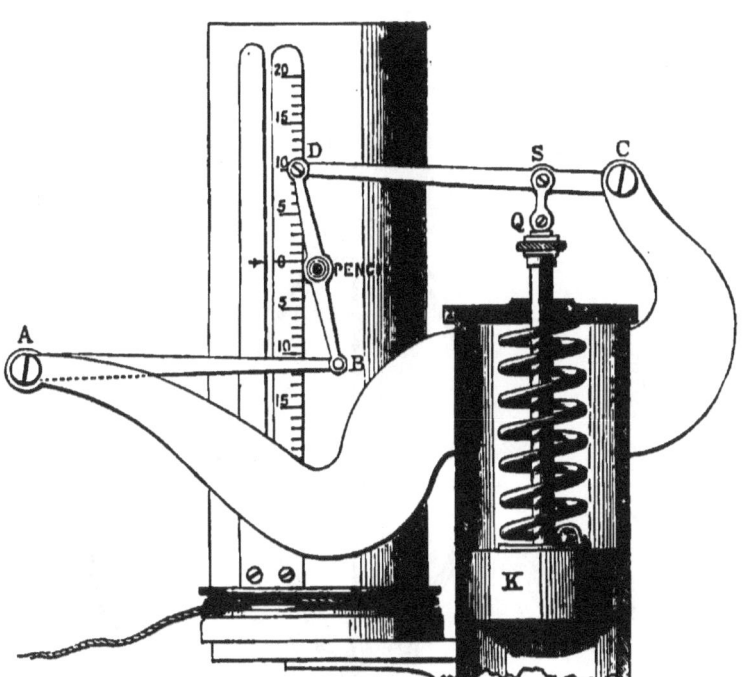

The parallel motion bars A B and C D carry the pencil, which traces out upon a drum a copy of the vertical movement of the piston K of the indicator, but magnified by reason of the attachment of the piston to a point s near the fulcrum of the bar C D. The principle of the apparatus is precisely the same as that which we have already explained, and the only difference consists in the application of the parallel motion bars to enlarge the diagram.

The drum derives its motion from any part of the engine whose movement is coincident with that of the piston, and the spiral spring can be changed so as to suit different engines. The connecting link is not set perpendicularly to the bars A B, C D, but makes an angle with them as shown; an artifice which causes the pencil to describe a line free from any sensible curvature.

136. The *Governor of a steam-engine* usually appears under the form invented by Watt, and has proved of the greatest possible value in steam machinery.

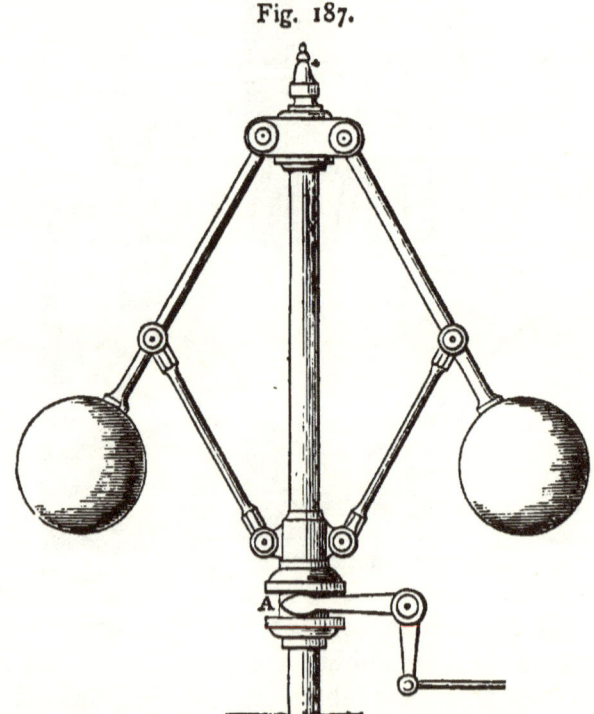

Fig. 187.

The diagram shows the construction of this well-known piece of apparatus, and the principle of its action is briefly the following :—

Watt's Governor.

The engine imparts rotation to the balls of a heavy conical pendulum, and maintains them at a certain inclination to the vertical; if the velocity of the engine be increased, the balls open out more widely; if it be diminished they collapse, and in doing so they move the end, A, of a system of levers which is connected with a throttle valve, and thereby regulate the supply of steam to the cylinder.

An engineer can easily arrange that the variation in speed admitted by the governor shall not exceed one-tenth of the mean velocity, but it is of the essence of the invention that some change in the speed should be admissible: the balls cannot alter their position unless the time of a revolution changes, and they cannot accumulate such additional momentum as may be sufficient to move the valve until the rate of the engine has sensibly altered.

In some cases, as where the engine drives machinery for very fine spinning, it may be desirable to obtain an almost absolute uniformity of motion; or, again, it may be an object to avoid the fluctuations in speed to which the common governor is liable when any sudden change occurs in the load upon the engine.

137. In order to control the engine with almost theoretical exactness, and to provide against the objections to which Watt's governor is exposed in certain extreme cases, Mr. Siemens has put forward a remarkable adaptation of epicyclic trains to the conical pendulum; and we shall proceed to an examination of his invention.

The original construction of this governor is exhibited in the diagram, and is better adapted for the purposes of explanation than a more recent arrangement. (Fig. 188.)

An epicyclic train of three equal wheels, A, B, C, is placed between the driving power and the conical pendulum; of these, A is driven by the engine, C is connected with the pendulum, and B is capable of running round A and C to a small extent defined by stops, the joints at F and B being so constructed as to permit of such a motion.

Fig. 188.

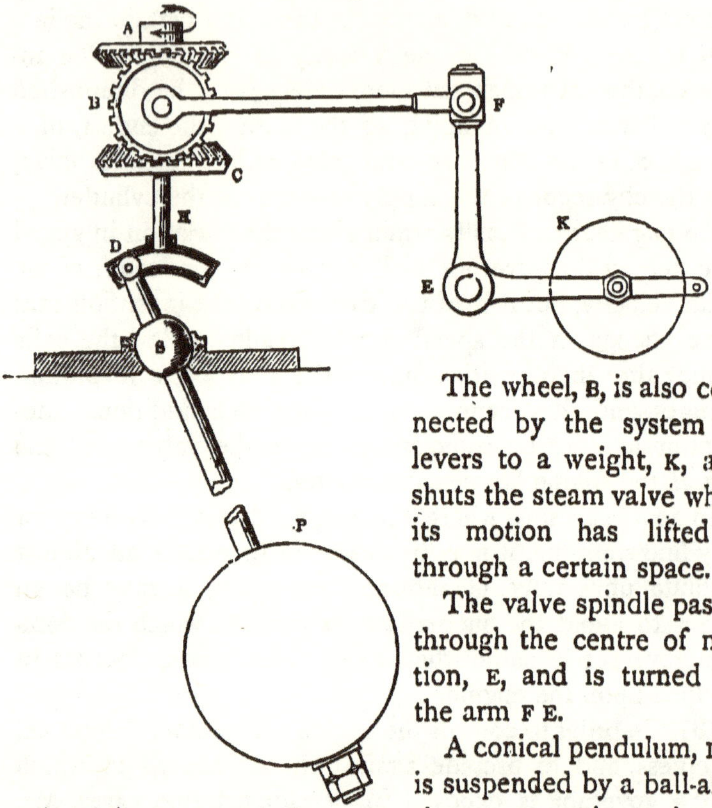

The wheel, B, is also connected by the system of levers to a weight, K, and shuts the steam valve when its motion has lifted K through a certain space.

The valve spindle passes through the centre of motion, E, and is turned by the arm F E.

A conical pendulum, D P, is suspended by a ball-and-socket joint at S, and the extremity D moves in a circular groove, D H. In this way the rotation of C is communicated directly to the pendulum.

It will be seen that a certain amount of maintaining force is absorbed in preserving the pendulum at a constant angle with the vertical, and it is a part of the contrivance to increase artificially the friction which opposes the motion of the pendulum, and thus finally to make the pressure exerted by the weight, K, an actual measure of the amount of such maintaining force.

The governor is at work when the velocity of the engine is just sufficient to keep K raised through a small space.

In order to understand the peculiar action introduced by

the epicyclic train, we should remember that one of these two things will happen: either A and C will turn at the same rate, or else B will shift its position and run round the axis A H; there can be no departure from the rigid exactness of this statement.

Now the wheel C is connected with the pendulum, and its rotation cannot be maintained without a constant expenditure of force; in other words, the tendency of C is to lag behind A, and to cause B to run round the axis A H.

This indisposition in C to accept the full velocity of A is artificially increased by the friction until B shifts its position and raises the weight K permanently, and then of course it follows that the pull of K evidences itself as a constant pressure tending to drive the wheel C.

The pendulum being in this manner retained in permanent rotation, suppose that any increase were to occur in the velocity of A; the wheel C is in connection with a heavy revolving body, and can only change its velocity gradually, but K is already lifted, in the sense of being counterpoised, and the smallest increase of lifting power can therefore raise it higher; thus the tendency to an increase in the velocity of A will at once cause B to change its position, and will control the steam valve.

So sensitive is this form of governor to fluctuations in speed, that an alteration of $\frac{1}{50}$th of a revolution may suffice to close the throttle valve altogether. It is in its power to move the valve, as well as in its sensitiveness, that this arrangement presents so remarkable a contrast to Watt's governor, where the moving force on the valve spindle is only the difference between the momenta stored up in the two positions of the balls. In this form of governor the power is only limited by the strength of the rods and levers; for it is apparent that the whole momentum stored up in the revolving pendulum would in an instant be brought to bear upon the valve spindle if any sudden alteration were to occur in the velocity of the wheel A.

In the method which has been adopted at Greenwich for registering the times of transits of the stars by completing a galvanic circuit at the instant of observation, a drum carrying a sheet of paper is made to revolve once in two minutes. A pricker actuated by an electro-magnet, and moving slowly in a lateral direction, is set in motion at the end of each beat of the seconds' pendulum of a clock, and thereby makes a succession of punctures in a spiral thread running round the drum. The observer touches a spring at the estimated instant of the time of transit of a star across a wire of the telescope, and, producing a puncture intermediate to those caused by the pendulum, does in fact record the exact period of the observation. The regularity of motion in the drum is a matter of vital importance, and is ensured by the employment of a clock train moving under the control of this pendulum of Mr. Siemens.

138. A further illustration of aggregate motion occurs in machinery for drilling and boring.

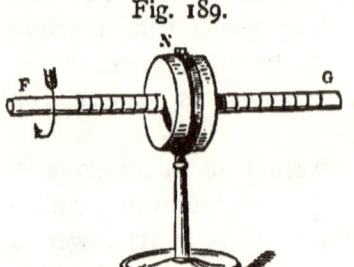

Fig. 189.

In a drilling machine the spindle which carries the cutting tool revolves rapidly, and at the same time advances slowly in the direction of its length.

The movement is obtained upon a very obvious principle. Conceive a nut, N, to be placed upon a screw-bolt, F G, and to be so held in a ring or collar that it can rotate freely without being capable of any other motion.

If the nut be fixed, and F G be turned in the direction of the arrow, it is clear that the bolt must advance through the nut; if, again, the screw be prevented from turning, and the nut be made to rotate in the same direction as before, the bolt will come back again. And, finally, if by any contrivance different amounts of rotation be impressed at the same time upon the nut and the screw, the bolt will receive the two longitudinal movements simultaneously, and the aggregate motion will be the sum or difference of these component parts.

139. Suppose the wheels D and C to be attached to the bolt and nut respectively, and to be driven by the pinions A and B, which are fixed upon the same spindle; and let A, B, C, D represent the numbers of teeth upon the respective wheels.

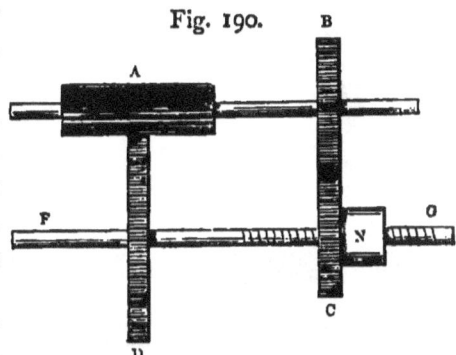

Fig. 190.

If (a) be the number of rotations made by either A or B while the nut fixed to C makes m rotations, and the wheel D makes n rotations,

we shall have $\dfrac{m}{a} = \dfrac{B}{C}$

and $\dfrac{n}{a} = \dfrac{A}{D}$.

Therefore (a) rotations of A will cause a travel of the bolt F G through a space

$$= (m - n) \times \text{pitch of the screw}$$
$$= a \left(\frac{B}{C} - \frac{A}{D} \right) \times \text{pitch of the screw.}$$

140. We shall proceed to examine the construction of a small *Drilling Machine*, which may be worked either by hand or by steam-power, but is *not self-acting*.

The general arrangement of the machine is shown in Fig. 192, the power is applied to turn the bevel wheel D, which again drives C, and causes the case or pipe containing the drill spindle to rotate. This provides for one part of the motion, viz. the rotating of the drill spindle, and the hand wheel K drives the spur wheels M and N, and advances the drill into the work in a manner which we shall endeavour to make clear.

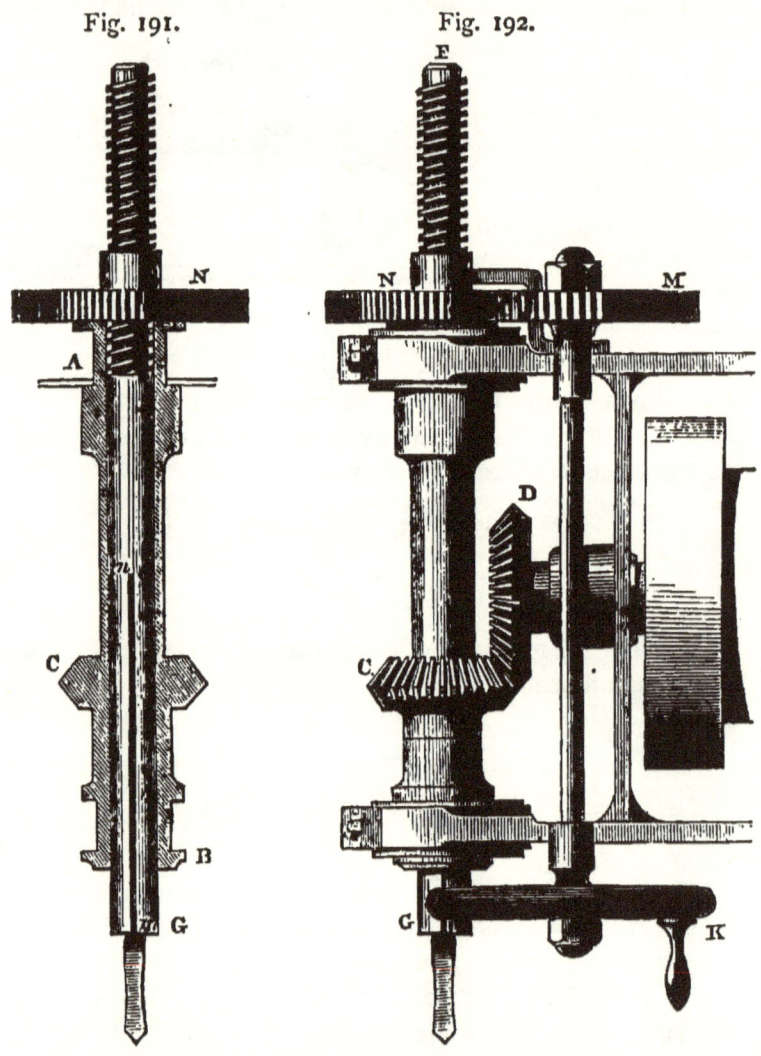

Fig. 191.   Fig. 192.

The drill spindle is formed in two pieces, as shown in Fig. 191, and the upper or screwed portion does not rotate with the lower cylindrical portion which carries the drill, but simply moves it up and down by means of a collar without interfering with its rotation. The screwed piece works in a nut forming the boss of the wheel N, and is prevented from rotating by a feather sliding in a groove or slot which runs along the whole length of the screw, and which cannot be seen in the view given in the drawing; the feather itself being fixed in a stop-collar at N.

Hence the rotation of the wheel N, by reason of its connection with the hand wheel K, will raise or depress the whole spindle as required.

The rotation of the drill spindle is provided for by cutting a groove *m n* in the lower part of it, and attaching a corresponding projection or feather to the inside of the pipe A B, this allows the spindle to move lengthways in the pipe and ensures its rotation just as if it were a part of the tube in which it is held.

A machine of this construction might easily be made self-acting as in a very useful form manufactured by Messrs. Smith, Beacock, and Tannett. Here the screwed spindle is not employed, but a rack and pinion is substituted for it, and the pinion is slowly raised or depressed by an endless screw and worm wheel set in motion by a hand wheel similar to K.

The self-acting portion consists of a small cone pulley, which draws off a motion of rotation from the driving shaft, and the axis of this pulley is fitted with a second endless screw and worm wheel placed just over the hand wheel, and which can be slid into gear so as to produce the self-acting motion.

Thus the same slow rotation may be given to the driving pinion on the axis of the hand wheel by the steam-power, which is otherwise given to it directly by the workman; the cone pulley of course providing for varying amounts of feed according to the requirements of the work.

141. A Drilling Machine by Mr. Bodmer, of Manchester, is made self-acting in the following manner:—

The drill spindle (Fig. 193) has a screw-thread traced upon it; a groove is cut longitudinally along the spindle, and a projection upon the interior of the boss of the wheel D fits accurately into the groove.

Thus the spindle can traverse through the wheel D, although the spindle and wheel must turn together.

Fig. 193.   Fig. 194.

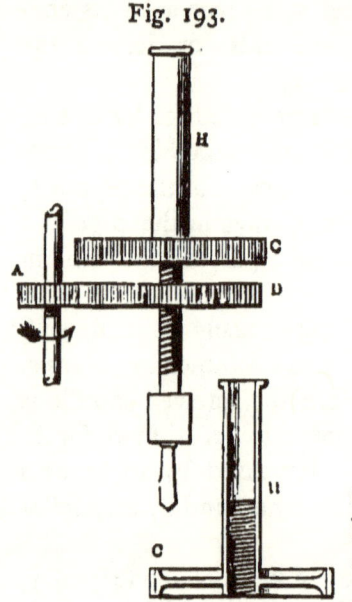
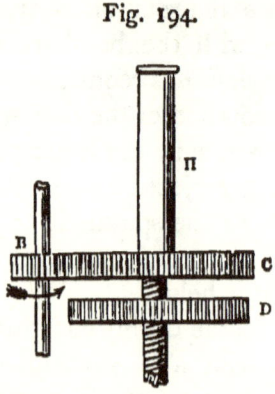

A nut H, in the form of a pipe, and having a wheel, C, at the bottom of it, receives the spindle; this wheel and pipe are shown separately in section.

If a pinion, A, turning in the direction shown by the arrow, engage the wheel D, it will screw the spindle rapidly out of the pipe H, and bring it down towards the work.

Suppose a second pinion, B, turning in the same direction as A, to act upon C, it will move the nut instead of the screw, and the drill spindle will rise rapidly so long as it is prevented from rotating. (Fig. 194.)

Thus far we have provided for bringing the spindle down to its work, and for raising it up again; it remains to apply the principle of aggregate motion, and to cause the drill spindle to become the recipient of these two movements in a nearly equal degree, and thereby to ensure the slow descent accompanied by a rapid rotation, which is required in process of drilling.

The result of the combination is shown in Fig. 195, where the wheels A and B are moved together; the wheel A tends to depress the spindle, the wheel B tends to raise it, and the spindle descends by the difference of these two motions, having further the motion of rotation given by the wheel A.

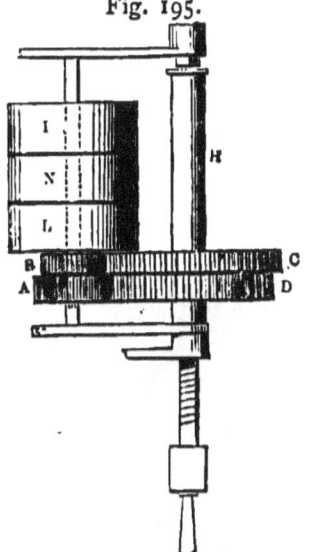

Fig. 195.

The motions of A and B are obtained from the driving pulleys I, N, and L.

I is an idle pulley; N drives A, and L drives B. When the strap is on N the drill descends to the work, when the strap is on L it ascends from the work; and when the strap is partly on N and partly on L the drilling proceeds.

The practical objection to this movement is that the rate of feed is invariable so long as the train of wheels remains the same; it is much better to control the feed by means of a cone pulley, where the strap can be readily shifted so as to change the advance of the cutter.

142. Sir J. Whitworth's *Friction Drilling Machine* is an elegant application of the principle under discussion.

A D is the drill spindle, which is driven in the usual manner by the bevil wheel B.

E and F are two worm wheels embracing the screwed portion of the spindle upon opposite sides; they are of peculiar construction, being hollowed out so as to fit against the small screwed spindle, and they work with a V-threaded screw upon A D.

If E and F be prevented from turning, they will form a nut through which the spindle will screw itself rapidly.

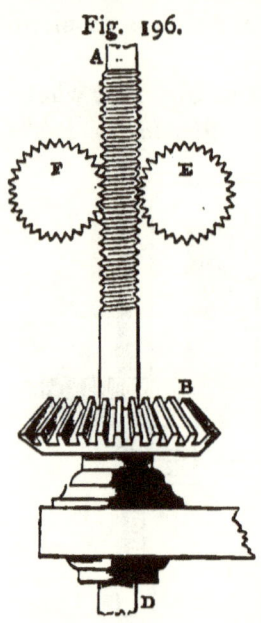

Fig. 196.

If E and F be allowed to turn quite freely, the drill spindle will set them in motion, and the nut will be virtually eliminated. The drill spindle may then be regarded as the recipient of two equal and opposite motions; it is depressed by screwing through the nut, it is elevated by the turning of the wheels.

If the rotation of the wheels be in any degree checked by the application of friction, the equality is destroyed, and the drill spindle descends to a corresponding extent.

A friction brake, regulated by a screw, restrains the motion of E and F, and gives a perfect command over the working of the machine.

When B is at rest the worm wheels act upon the screwed part of the spindle just as a pinion does upon a rack, and the drill can be rapidly brought down to the work.

This method of converting a screw and worm wheel into a rack and pinion is quite worthy of attentive consideration; it is employed in the well-known lathes by the same firm.

143. A *Boring Machine* would be employed to give an accurate cylindrical form to the interior surface of a steam cylinder.

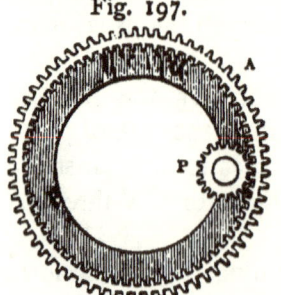

Fig. 197.

In the annexed example the boring cutters are attached to a frame which rides upon a massive cast-iron shaft or boring bar, and rotates with it; this frame is further the recipient of a slow longitudinal movement given by a screw.

An annular wheel, A, shaped as in the figure, rides loose upon the bar, and drives a pinion, P, at the end of the feeding screw which advances the cutters, the boring bar being recessed in order to receive the screw.

It is quite apparent that as long as the rotation of the wheel A is identical with that of the boring bar, the pinion P will not turn at all; and further, that a slow motion will be impressed upon P, if the rotation of A be made to lag a little behind that of the bar.

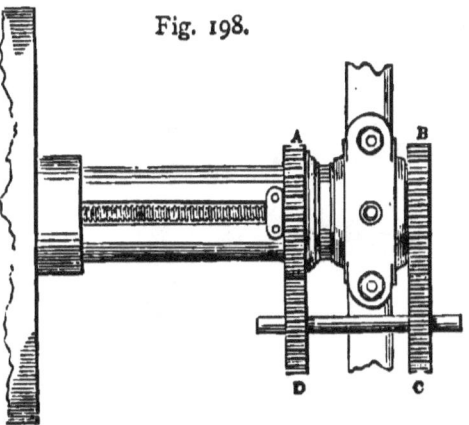

Fig. 198.

A spur wheel, B, is keyed to the bar, a small shaft fixed at the side carries the wheels C and D, and thus motion is imparted to A, the driver of the feeding screw. Let the numbers of teeth upon B, C, D be 64, 36, 35, and let the wheel A have 64 teeth, both upon the outside and the inside of its circumference, the pitch of the screw being $\frac{1}{2}$ an inch, and the number of teeth upon the pinion being 16.

$$\text{Here } e = \frac{B \times D}{C \times A}$$

$$= \frac{64 \times 35}{36 \times 64}$$

$$= \frac{35}{36}.$$

That is, A loses $\frac{1}{36}$th of a revolution for every complete rotation of the boring bar.

At the same time the pinion P moves through $\frac{1}{36} \times \frac{64}{16}$ or $\frac{1}{9}$ of a revolution, and the cutter advances through $\frac{1}{9} \times \frac{1}{2}$ an inch or through $\frac{1}{18}$th of an inch.

144. This slow rotation of the screw which advances the boring head may be obtained in a more simple manner by a combination which virtually embodies the sun and planet wheels of Watt.

Conceive that two wheels, A and B, of 40 and 80 teeth respectively, are attached to the bar C A B which has a centre of motion at C.

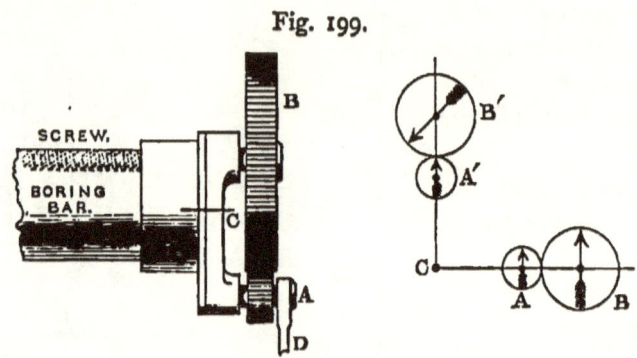

Fig. 199.

If the bar be carried round C, and A be made a dead wheel, the effect of depriving A of the rotation due to its connection with the arm will be to cause B to rotate relatively to the arm just as if the axes of both wheels were fixed in space.

The movement is shown in the diagram, where A has turned through half a right angle from its first position *relatively to the arm*, while the arm itself has been carried through a right angle.

The student will distinguish between the *absolute* and *relative* rotations of B; the absolute amount of the rotation of B is one right angle and a half.

This also appears from the formula, viz.

$$e = \frac{n-a}{m-a}.$$

Substituting the values $m = 0$, $e = -\tfrac{1}{2}$

we have $-\tfrac{1}{2} = \dfrac{n-a}{-a}.$

$$\therefore n - a = \frac{a}{2}$$

But $(n-a)$ represents the number of rotations of the wheel B relatively to the arm while the latter is making (a) revolutions, and the analysis therefore shows that the angular velocity of B relatively to the arm is half that of the arm itself, and also that both rotations take place in the same direction.

Further, it must be noted that the position of C makes no difference in the result, which will be quite the same if the point C be somewhere between A and B.

In the application of this movement to the boring machine, the centre of motion is between the axes of the wheels, in the line marked C in the diagram, and the numerical value of e is less than $\frac{1}{2}$, probably about $\frac{1}{5}$.

The wheel B is placed upon the axis of the screw which advances the boring cutters, the rotating arm being now a part of the solid end of the boring bar; the wheel A rides upon a separate stud, and is attached to a bar A D of some convenient length which passes through and rests upon a fork in an independent upright support placed at some little distance from the machine.

As the wheel A is carried round the axis of the boring bar this rod slides a little to and fro in the fork, and controls the wheel A so as to render it impossible for it to rotate, or, in other words, to make it a dead wheel.

The wheel B will now turn slowly under the action of A so far as its position relatively to the boring bar is concerned, and upon our supposition, the screw will advance the boring cutters by a space equal to its pitch in five complete revolutions.

This would give a feed dependent upon the pitch of the screw, which could of course be varied at once by changing the wheels A and B.

145. The *oval chuck* affords another example of aggregate motion. It is based upon the following property of an ellipse, which is taken advantage of in constructing elliptic compasses for drawing the curve.

P

Let A C A′, B C B′, represent two grooves at right angles to each other, and traced upon a plane surface; P D E, a rod furnished with pins at D and E: if this rod be moved into every possible position which it can assume while the pins remain in the grooves, the point P will describe an ellipse.

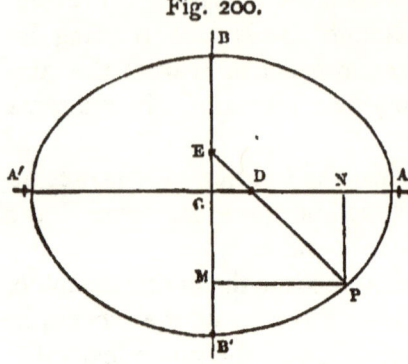

Fig. 200.

Draw P N $\perp^r$ to A C, and P M $\perp^r$ to C B′.

Let C N $= x$, P E $= a$
P N $= y$, P D $= b$.

Then $\dfrac{x}{a} = \dfrac{P M}{P E}$, and $\dfrac{y}{b} = \dfrac{P N}{P D} = \dfrac{E M}{P E}$

$\therefore \dfrac{x^2}{a^2} + \dfrac{y^2}{b^2} = \dfrac{P M^2 + E M^2}{P E^2} = 1,$

which is the equation to an ellipse.

In drawing an ellipse we should fix the paper and move the rod over it, but in turning an ellipse in a lathe we should fix the describing tool and move the piece of wood or metal underneath it; thus the conditions of the problem become changed, and the construction is modified accordingly.

An equivalent for the grooves A C A′, B C B′ may be arrived at as follows:—

Describe a circle about E of radius larger than E D, and let two parallel bars, Q R, S T, be connected by a perpendicular link H K equal in length to the diameter of the circle, and thus form a rigid frame embracing the circle, and capable of moving round it.

As the frame moves round the circle we must provide that H K shall pass through D in every position as represented in Fig. 202; if we now draw D C parallel to Q R, and E C parallel to H K, it is easy to understand that the imaginary

Oval Chuck. 211

triangle D C E in Fig. 202 is exactly the same as the triangle D C E in Fig. 200, and exists throughout the motion; and that whereas we formerly moved the bar E P over a fixed plane and described an ellipse, so now we have arranged to obtain the same motion with a fixed bar and a moveable plane, and shall trace out precisely the same curve.

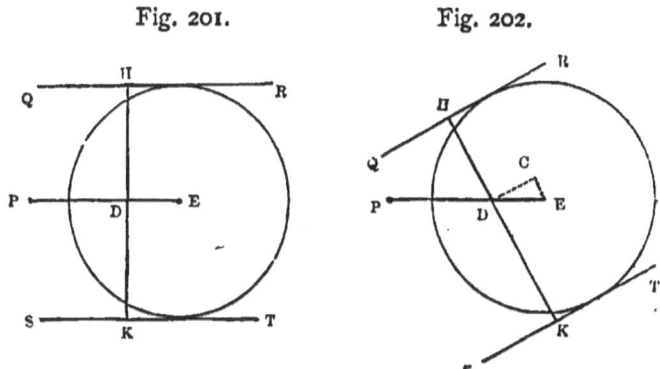

Fig. 201. Fig. 202.

This is a very good example of aggregate motion. The plane upon which the ellipse is traced is the subject of two simultaneous movements; by one of them a line, H K, in the plane is made to revolve round D as a centre, and by the other the same line receives a sliding motion in alternate directions through D.

Thus an oval, or more properly an ellipse, may be turned in the lathe.

CHAPTER VI.

MISCELLANEOUS CONTRIVANCES.

WE propose to examine in our concluding chapter various miscellaneous pieces of mechanism, and certain special contrivances which are of frequent occurrence in machinery, and with which a student of applied mechanics ought to render himself familiar.

146. *The fusee* is adopted in chronometers, and in most English watches, in order to maintain a uniform force upon the train of wheels, and to compensate for the decreasing power of the spring.

The spring is enclosed in a cylindrical barrel, and sets the wheels in motion by the aid of a cord or chain wound partly upon the barrel and partly upon a sort of tapering drum called a fusee.

As the spring uncoils in the barrel the pull of the cord decreases in intensity; at the same time, however, the cord

Fig. 203.

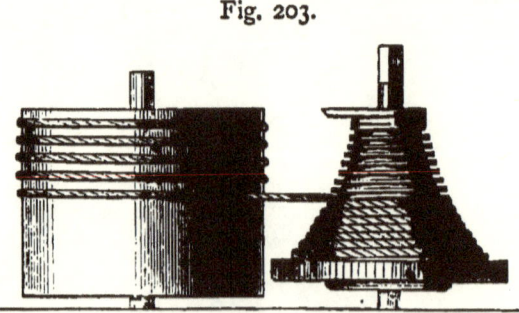

unwinds itself from the fusee, and continually exerts its

strain at a greater distance from the axis, that is, with a greater leverage, and with more effect.

147. The theoretical form of the fusee is an hyperbola, being the section of a right cone made by a plane parallel to the axis of the cone.

To prove this statement we must first recognise the law according to which an elastic body under extension or compression exerts a force of restitution whereby it tends to recover its original form.

This law was stated by Dr. Hooke as being contained in the maxim 'ut tensio sic vis,' by which it is intended to convey that when a body is extended or compressed in a degree less than that which produces a permanent derangement of form, the force necessary to keep it extended or compressed is proportional to such extension or compression.

Take a spiral steel spring balance, for example, hang upon it successive weights of 1, (2), (2 + 1), (3 + 1) lbs., the index point will descend through equal spaces for each additional pound weight, and will rise by equal spaces as each pound is successively removed.

Assuming the law to hold exactly when the spiral spring is subject to a force of torsion instead of one of direct extension, we shall have the pull of the spring proportional to the angle through which the barrel has been made to turn.

Let D P B A represent one-half of the section of a fusee, D P B being the curve whose equation is to be found.

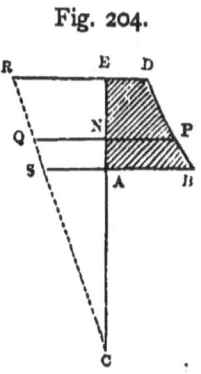

Fig. 204.

Draw D E, P N, B A perpendiculars on E A; take E R, Q N, S A to represent the pull of the spring when the chain is at the points D, P, and B respectively.

According to Hooke's law the force of the spring will decrease uniformly as the chain passes from D to B, therefore R Q S must be a straight line inclined to E A, produce it to meet E A in C.

Then $\dfrac{QN}{CN} = \dfrac{SA}{CA}$ which is a constant ratio, by reason of the law of elasticity.

Assume that this ratio is represented by m,
$$\therefore QN = m \cdot CN.$$

In order that the fusee may accomplish its object, the product of the pull of the spring into the arm NP must remain constant for every position of P.

Hence calling $CN = x$, $NP = y$ we have

(pull of spring) $\times NP = m \cdot CN \times NP = m\,x\,y$.

But this product is not to vary,
$$\therefore m\,x\,y = \text{a constant quantity,}$$
$$\text{or } x\,y = \text{a constant,}$$
which is the equation to an hyperbola.

In practice, where great accuracy is required, the strength of the spring is tested by fixing a light lever to the winding square of the fusee and observing whether the pull of the spring is balanced in every position by the same weight hung at the end of the lever. The fusee would be cut away a little where it was necessary to do so.

148. In mechanism the fusee is frequently employed to transmit motion instead of to equalise force, and enables us to derive a continually increasing or decreasing circular motion from the uniform rotation of a driving shaft.

The groove of the fusee may be traced upon a cone or other tapering surface, or it may be compressed into a flat spiral curve: in all cases the effect produced will be that due to a succession of arms which radiate in perpendicular directions from a fixed axis, and continually increase or decrease in length.

The fusee can of course only make a limited number of turns in one direction.

149. A *flat spiral fusee* occurs in spinning machinery, and serves to regulate the velocity of the spindles and to ensure the due winding of the thread in a succession of conical layers upon a bobbin or *cop*.

The Fusee.

The formation of the *cop* is a problem upon which a vast amount of mechanical ingenuity has been expended; and, without entering too much into details, we may observe that there are two distinct stages in the process of winding the yarn upon a spindle so as to produce a finished cop.

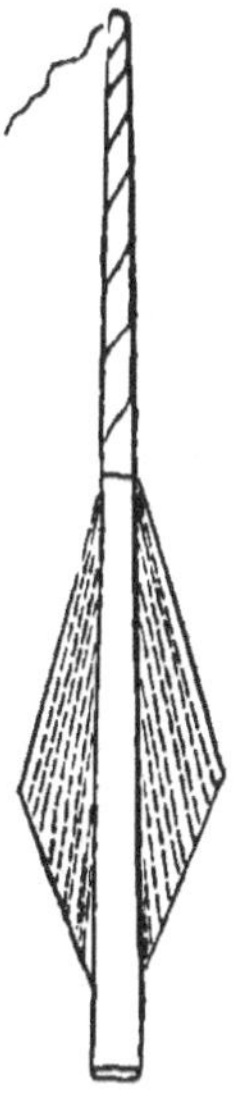

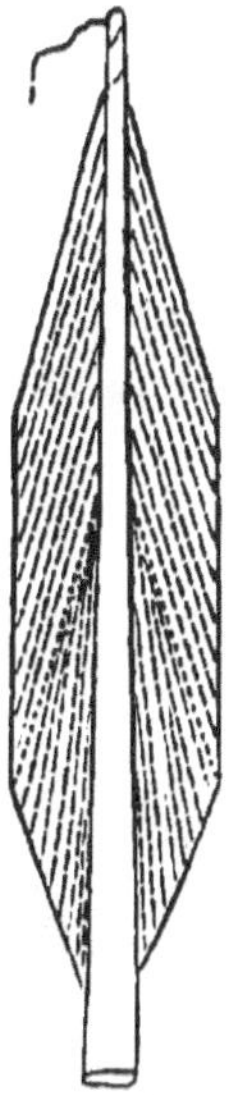

The *copbottom* (Fig. 205) is first formed upon a bare spindle by superposing a series of conical layers with a continually increasing vertical angle.

The body of the cop is then built up by winding the yarn in a series of equal conical layers. (Fig. 206.)

The winding-on of the yarn begins at the base of the cone, and proceeds upwards to the vertex; the spindles are driven by a drum which rotates under the pull of a chain, and it is evident that they can be made to revolve with increasing rapidity by placing a fusee upon the driving shaft and causing the chain to coil upon it.

Fig. 205.   Fig. 206.

Such an arrangement, as shown in Fig. 207, will be adapted to the winding of a uniform supply of thread upon a conical surface; and

Fig. 207.

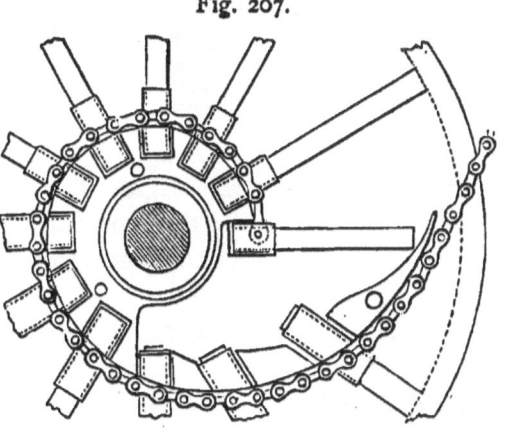

we can easily comprehend that a fusee of fixed dimensions will do very well for building up the body of the cop after the foundation is made. The main difficulty occurs in producing the copbottom, where the series of conical layers of continually increasing vertical angle demands a fusee whose dimensions shall gradually contract towards the centre.

The method of contracting the form of the fusee may be explained as follows:—

Fig. 208 represents portions of two flat discs having axes at A and B, and upon which are cut radial and curved grooves in the manner indicated; it being arranged that when one plate is placed upon the other, the pins P and Q shall travel in both sets of grooves at the same time.

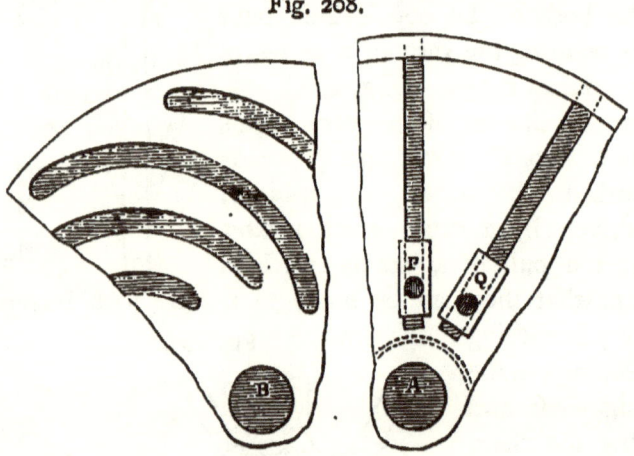

Fig. 208.

We can easily see that the blocks which carry the pins will move along the radial grooves as the disc B turns relatively to A, and that by this combination we can obtain a spiral fusee of any required form, and can contract or enlarge its dimensions at pleasure.

150. *Mr. Roberts's winding-on motion* reposes upon the principle of the fusee, though in a modified form.

Let one end of a rope which is coiled round a drum be attached to a point, P, in the moveable arm C P: it is evident that the rotation of C P about the centre of motion C will cause some portion of the rope to be unwound from the barrel. (Fig. 209.)

The Fusee.

Draw C S perpendicular to the direction of the rope; then, at any instant of the motion, the arrangement supplies the jointed rods, C P, B Q, mentioned in Art. 53, and it is manifest that the rate at which the string is unwound will vary as the perpendicular C S.

This rate is greatest when C P is perpendicular to P Q, but decreases to nothing when C S vanishes,

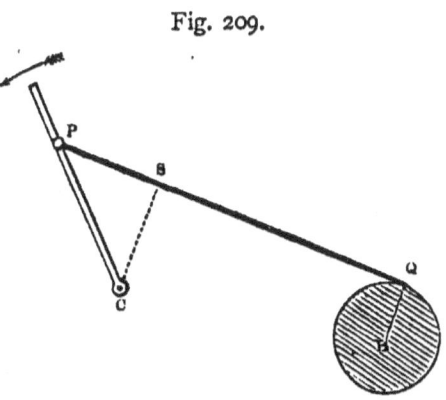

Fig. 209.

and here, therefore, the varying arm of the fusee exists in a latent form.

Next conceive that the conditions are changed, and that the drum B moves to the right hand through a moderate space while C P remains fixed; the cord will unwind from the drum with a nearly uniform velocity.

If, finally, the arm C P be not fixed, but be made to move from a position a little to the left of the vertical into one nearly horizontal during one journey of the drum, it is abundantly clear that we shall subtract from the uniform

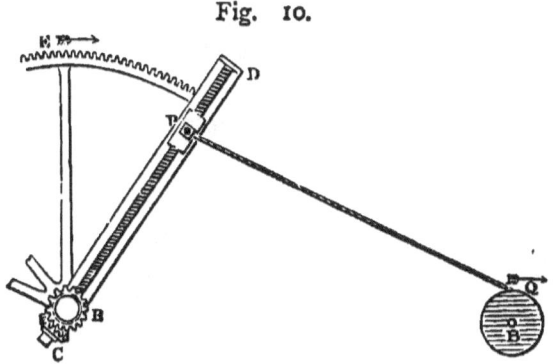

Fig. 10.

motion of unwinding that amount which is due to the action

of a fusee, and that if the spindles derive their motion from the rotation of the drum they will continually accelerate as the drum recedes from C P. In this way we can make up the body of the cop. To form the copbottom, it is necessary that the winding on should begin more rapidly, and should gradually diminish; this character of motion is produced by causing the nut P to traverse C D in successive steps during each journey of the drum. As soon as the cop has attained its full diameter, the nut ceases to travel along C D, and the thread is wound in uniform conical layers. (Fig. 210.)

151. If two cords be wound in opposite directions round a drum, A, and the ends of the cords be fastened to a move-

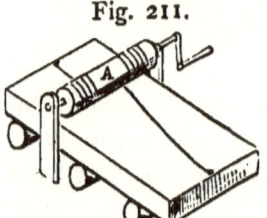

Fig. 211.

able carriage, it is evident that the rotation of A in alternate directions will cause a reciprocating movement in the carriage.

This is a mangle in its simplest form, and the objection that the handle must be continually turned in opposite directions may be obviated by the use of the mangle wheel.

It is clear that if the drum were divided into two portions, and that each half instead of being cylindrical were formed into a fusee, the motion of the piece driven by the rope would be no longer uniform but would vary with the dimensions of the fusee.

Hence the drum A has been replaced by a spiral fusee in

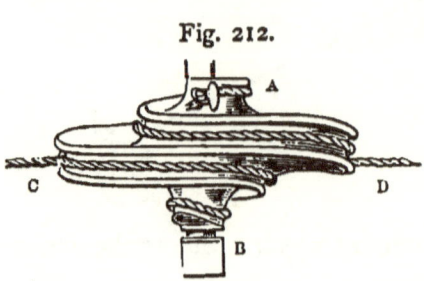

Fig. 212.

the self-acting mule of Mr. Roberts, and thus the motion of the carriage is gradually accelerated until it has reached the middle of its path, and then decreases to the end of the movement.

The Snail. 219

It must be understood that the cord fastened at A goes off at C, while that fastened at B passes on to D.

152. A helical screw of a varying pitch traced upon a cylinder would produce a similar variable motion of the

Fig. 213.

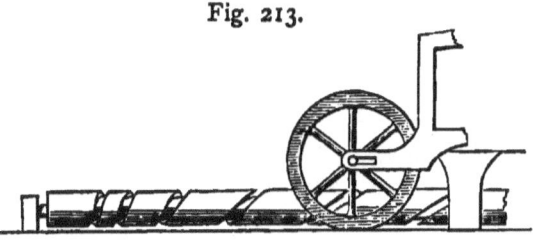

mule-carriage, and has been applied in a machine constructed upon a different principle, in order to obtain a continually decreasing motion of the carriage. It replaces the fusee.

153. The *snail* is chiefly found in the striking part of repeating clocks. It is a species of fusee, and is used to define the amount of angular deviation of a bent lever A B D, furnished at the end A with a pin which is pressed against the curve of the snail by a spring, and is attached at the other end to a curved rack, whose position determines the number of blows which will be struck upon the bell.

Fig. 214.

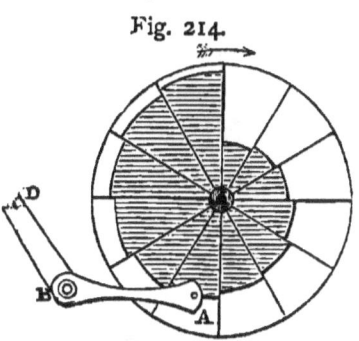

In order to form the snail, a circle is divided into a number of equal parts (twelve, for example), and a series of steps are formed by cutting away the plate and leaving a circular boundary in each position.

As the snail revolves, A B D passes by jerks into twelve different positions, and the clock strikes the successive hours.

Since the point A describes a circle about B, it is clear that the depth of each step must vary in order to obtain a constant amount of angular motion in the arm B A during each progressive movement. It will be seen that the circular arc described by the end of the small lever has its tangent at A, when in the position sketched, parallel to the vertical diameter which divides the snail into two equal parts, and this reduces the inequality between the steps.

154. The *disc and roller* is equivalent to the fusee, and is now but little used, on account of the probability that the roller will occasionally slip.

This arrangement consists of a disc A, revolving round an axis perpendicular to its plane and giving motion to a rolling plate B, fixed upon an axis which intersects at right angles the axis of the disc A.

Fig. 215.

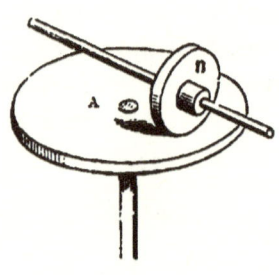
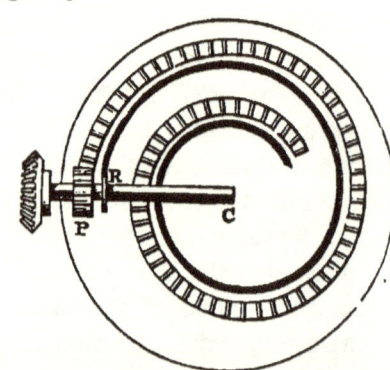

Supposing the rotation of the disc to be uniform, that of the roller B will continually decrease as it is shifted towards the centre of A, and conversely.

This is precisely the effect produced by a fusee.

The roller may be a wheel furnished with teeth, and may roll upon a spiral rack as shown in the diagram.

As the disc revolves, the pinion P slides upon the square shaft, and is kept upon the rack by the action of a guide-roller, R, which travels along the spiral shaded groove.

This example is by no means put forward as a good mechanical contrivance, for indeed the disc and roller possesses an inherent defect which should be diminished as much as possible in practice and not exaggerated. The bounding circles of the roller run with the *same* linear velocity, whereas the circular paths upon which they are both respectively supposed to roll move with *different* linear velocities, by reason of their being concentric circles of unequal size traced upon a plate which rotates with a uniform angular velocity about an axis through the common centre.

It is geometrically impossible that the bounding circles of the roller can both roll together, or the combination will fail in exactness as a piece of pure mechanism, and thus two rollers running round upon a flat surface or bed stone in the manner suggested form an excellent pulverising or grinding apparatus. These rollers are called edge-runners: they are of large size and very weighty, and are placed near to the vertical axis about which they run.

155. Where a train of wheels is set in motion by a spring enclosed in a barrel, it becomes of consequence not to overwind the spring; the *Geneva stop* has been contrived with the view of preventing such an occurrence, and will be found in all watches which have not a fusee.

Here a disc A, furnished with one projecting tooth P, is fixed upon the axis of the barrel containing the mainspring, and is turned by the key of the watch.

Another disc, B, shaped as in the drawing, is also fitted to the cover of the barrel, and is turned onward in one direction through a definite angle every time that the tooth P passes through one of its openings, being locked or prevented from moving at other times by the action of the convex surface of the disc A.

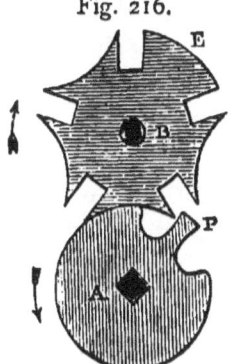

Fig. 216.

In this manner each rotation of A will advance B through a certain space, and the motion will continue until the convex surface of A meets the convex portion E, which is allowed to remain upon the disc B in order to stop the winding up.

The winding action having ceased, the discs will return to their normal positions as the mechanism runs down.

Instead of supposing A to make complete revolutions let it oscillate to and fro through somewhat more than a right angle; then B will oscillate in like manner and will be held firmly by the opposition of the convex to the concave surface except during the time that P is moving in the notch.

This movement is employed in the reversing motion alluded to in Art. 65; the pulley which carries the driving cord is fixed on the axis of B, and is made to reciprocate by the action of a driving pin attached to a small lever weighted at one end and centred upon the axis A. The table pushes this lever over and it falls either to the right or the left; in doing so the pin enters an opening in B and causes the first pulley in the series to reciprocate, while the form of the stop prevents any movement except at the end of each travel of the table, and the cutter is held firmly in the right position.

156. *The Double Eccentric for reversing an Engine.* When the piston is near the middle of its stroke in a locomotive or marine engine, the slide-valve will have moved over the steam-ports in the manner pointed out in Fig. 217.

The slide-valve is connected with a point in the circumference of a small circle which represents the path of the centre of the eccentric pulley, and the piston is connected with a point in a larger circle, representing the path of the centre of the crank-pin.

The piston and valve are shown as separated in the drawing, but the small circle is repeated in the position which it actually occupies, and the method of reversal is the following:—

In the upper diagram the piston is supposed to be moving to the right, and the valve to the left, the piston having travelled so far in its stroke that the valve is returning to cut off the steam; in order, therefore, to change the motion, we must drive the piston back by admitting the steam upon the opposite side, and by letting out that portion of the

The Double Eccentric. 223

steam which is urging the piston forward. Hence we must move the valve into the position shown in the lower diagram,

Fig. 217.

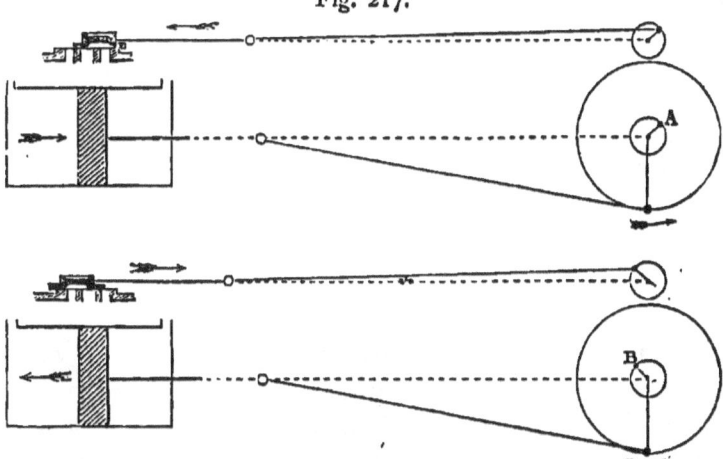

and shift the centre of the eccentric pulley from A to B; the piston will then return before it reaches the end of the cylinder, and the movement of the engine will be reversed.

In examining the diagram it should be understood that the crank which works the slide-rod is inclined at an angle somewhat greater than 90° to the crank which is attached to the piston; and also that the crank of the slide-rod is always in advance of the larger one in its journey round. The engine would not work if the larger crank were to turn in the opposite way to that shown in the sketch.

This explanation shows that in reversing an engine we must either shift the eccentric from the one position into the other, or else we must employ two eccentrics, and provide some means of connecting each of them in turn with the slide-valve.

The problem, as presented in the latter alternative, has been solved in a most convenient and elegant manner by an arrangement commonly known as *Stephenson's link motion*.

Here A B is the starting lever, under the control of the engine driver, and is represented as being pushed forward in the direction in which the engine is moving; C D is the *link*, provided with a groove, along which a pin can travel; a short lever, centred at R, is connected at one end, Q, with the slide-valve, and at the other end with the pin which moves in the link.

It is clear that so long as the pin remains near the point D, the lever centred at R will be caused to oscillate just as if the pin were attached to the extremity of the outer eccentric bar, and that the outer eccentric alone will be concerned in the motion of the valve.

If now the engine driver wishes to reverse his engine, he pulls back the lever A B, and by doing so he raises the link C D until the pin comes opposite to the end of the inner eccentric bar: the raising of the link is caused by the motion imparted to the bell-crank lever, G E F, which is centred at the point E. A counterpoise to the weight of the link is attached to the axis passing through E at some little distance behind the bell-crank F E G, so as to be out of the way of the moving parts, and the object of this counterpoise is to enable the engine driver to raise the heavy link and bars easily.

The inner eccentric bar now alone comes into play, and the two eccentrics being fastened to the crank axle at the angles indicated in the first part of the article, it is apparent that the valve will be shifted, and that the action of the engine will be reversed.

A special advantage of this link motion in practice is to be found in the power which it gives to the engineer of regulating the supply of steam admitted into the cylinder. By moving the starting lever into intermediate positions the amount of travel of the valve is reduced at pleasure; for it is evident that no steam at all can get into the cylinder when the lever is half-way between its extreme positions, and that varying amounts of opening of the steam-ports, increasing to the maximum value, will occur when the lever is pushed over by successive steps.

The Double Eccentric. 225

Fig. 218.

157. *Step wheels* constitute a modification of toothed wheels; they are due to Dr. Hooke, and are used to ensure a smooth action in certain combinations of wheel-work.

It is evident that the action of two wheels upon each other becomes more even and perfect when the number of teeth is increased, but that the teeth at the same time become weaker and less able to transmit great force.

Dr. Hooke's invention overcomes the difficulty, and virtually increases the number of teeth without diminishing their strength.

Several plates or wheels are laid upon one another so as to form one wheel, and the teeth of each succeeding plate are set a little on one side of the preceding one, it being provided that the last tooth of one group shall correspond within one step to the first tooth of the next group. The principal part of the action of two teeth occurs just as they pass the line of centres, and there are now three steps instead of one from the tooth A to the tooth B.

Fig. 219.

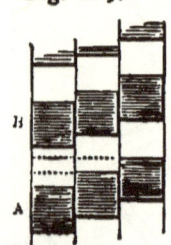

A single oblique line might replace the succession of steps, but we should then introduce a very objectionable endlong pressure upon the bearings of the wheels.

Pinions of this construction are to be met with in planing machines, and are employed to drive the rack which is underneath the table; so again step wheels are used in marine steam engines where the screw shaft is driven from an axis considerably above it. They are valuable where strength and smoothness of action are to be combined.

158. There is a curious movement derived from the employment of a dead wheel in a train which has been applied by Mr. Goodall in a machine used for grinding glass into powder by the action of a pestle and mortar. The pestle is made to sweep round in a series of nearly circular curves contracting to nothing, and then expanding again so as to command the whole surface of the mortar.

Variable Crank.

We shall show that the contrivance is merely a solution of the problem of obtaining an *expanding and contracting crank*.

Fig. 220.

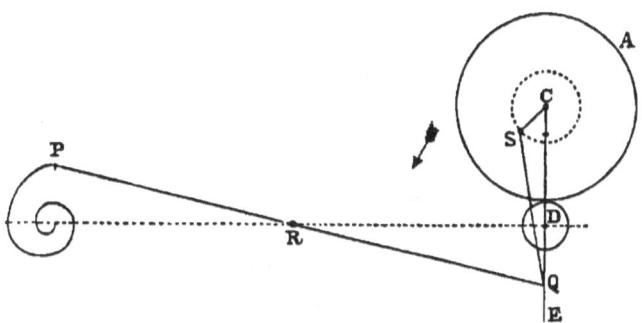

Let C Q E be a crank whose centre of motion is D; conceive that D is a dead wheel on the same axis, and that A is a larger wheel riding upon one end C of the crank arm.

Suppose, further, that a piece Q, capable of sliding along C E, is attached by a link S Q to a point S in the circle A which is not its centre; and, finally, that a pencil at P is connected with Q by a link P R Q constrained always to pass through a fixed point R.

As the circle A and the crank C Q travel together round the dead wheel D, it has been proved in Art. 144 that the wheel A will turn relatively to the arm just as it would do in an ordinary train with fixed axes. Hence the point S travels slowly round in the dotted circle, thereby causing the point Q to move to and fro along C Q.

It may be arranged that Q shall start from D, and it will travel along C E through a space equal to twice C S. When Q is at D, the point P is motionless, whereas, while Q moves further from D, and continually sweeps round, by virtue of its being a point in a revolving crank, it is evident that P will trace out an expanding spiral, which will return again to nothing when Q is pulled back to D by the action of the wheel A.

It now only remains for us to consider what would be the actual construction of the apparatus.

The drawing is taken from a model, and not from the machine itself.

Fig. 221.

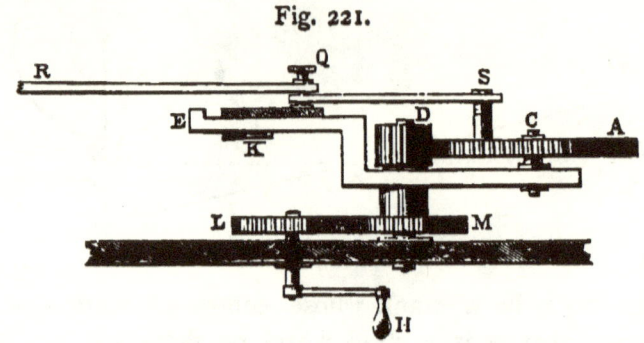

The crank C D E is a bar whose parallel arms are connected by a vertical piece, and which carries the wheel C upon one arm and the sliding piece K Q upon the other arm. This crank is driven by the spur wheels L and M connected with the handle H; it therefore rotates round the axis of the dead wheel D and carries C and K Q upon opposite sides of the vertical axis through D.

The link S Q connects the wheel C with the piece K Q, and this latter piece is again connected by Q R with the pestle, it being provided that Q R shall pass through a guide at some fixed point about half-way between Q and the mortar. The pestle is swung from a ball and socket joint at some convenient height above P.

The rotation of the crank round the dead wheel causes C to turn slowly upon its own axis, the point S therefore travels slowly round C, hence the end Q of the connecting rod Q R P is sometimes at a distance from D, and at other times is exactly over it, and during the whole time Q is a part of the crank D Q, and sweeps round with the arm. Thus the required motion is provided for.

Speed Pulleys.

159. *Speed pulleys* are so called because they allow of the transfer of different velocities of rotation from one shaft to another; they are much used in engineers' factories.

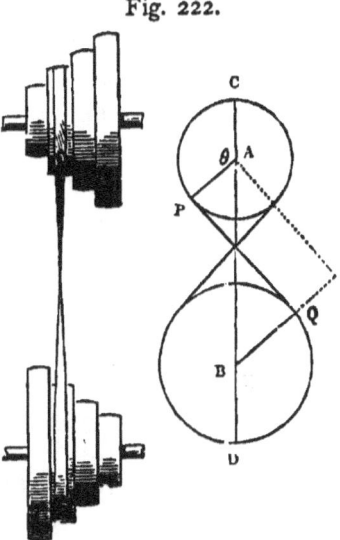

Fig. 222.

They are made in a series of steps, as shown in the diagram, one pulley being the counterpart of the other, but pointing in the opposite direction.

If the steps be equal, as is commonly the case, the sum of the radii of each pair of opposite pulleys will be a constant quantity.

It is a geometrical fact that when two circles are placed with their centres at a given distance, and are so related that the sum of their radii remains constant, an endless crossed band connecting both the circles will not vary in length in the smallest degree during the change in the actual diameter of each circle.

Hence a crossed strap will fit any pair of the pulleys in our series with perfect exactness.

The proof is the following :—

Let A and B represent two pulleys whose radii are A P and B Q, and assume that A P + B Q remains unchanged while A P and B Q respectively increase and diminish,

and let $AP = x$, $BQ = y$, $PAC = DBQ = \theta$.

Then $CPQD = x\theta + y\theta + PQ$
$= (x + y)\theta + PQ.$

But it is clear that if A P be increased by a given quantity, and B Q be diminished by the same quantity, we shall not change the length of P Q, by reason that the alteration will only cause P Q to move through a small space parallel to itself between the lines A P, B Q, which are also parallel.

Also $x + y$ is constant by hypothesis,

∴ C P Q D remains unaltered in length so long as our condition holds good.

It may be interesting to examine this matter a little further and to find an expression for the length of an open band connecting two given pulleys. We will assume that we are dealing with step pulleys, the sum of the radii being $2a$ in every case, and x being the depth of the step or steps, or the quantity by which either radius differs from the assumed value of the semi-sum of the radii.

Fig. 223.

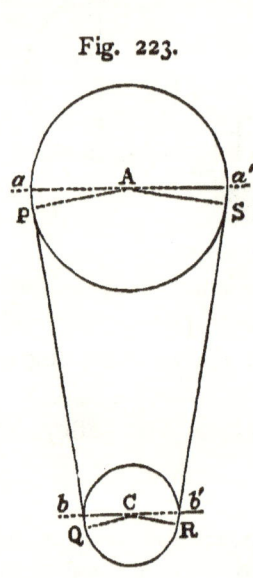

Let $AP = a + x$
$CQ = a - x$
$AC = d$
$PAa = \theta = QCb$

and let l = length of the band P Q R S. Then the curved portions of the band resting upon the pulleys are $(\pi + 2\theta)(a + x)$ and $(\pi - 2\theta)(a - x)$ respectively.

$$\therefore l = (\pi + 2\theta)(a + x) + PQ$$
$$+ (\pi - 2\theta)(a - x) + RS$$
$$= 2\pi a + 4\theta x + 2 PQ.$$

Now $\dfrac{PQ}{AC} = \cos\theta$

$\therefore PQ = AC \cos\theta = d \cos\theta$

Also $\dfrac{AP - CQ}{AC} = \sin\theta$

$\therefore 2x = AC \sin\theta = d \sin\theta$
$\therefore l = 2\pi a + 2d\theta \sin\theta + 2d \cos\theta.$

It is evident that l is no longer constant and that it must necessarily change when x or θ changes, still the variation of length may be so little as to be disregarded under the ordinary proportions occurring in a workshop.

Since θ would seldom represent an angle so large as 10°, and we have pointed out in Art. 126 how small a difference exists between sin θ and θ within even larger limits, we will assume that sin $\theta = \theta$,

then $\cos \theta = 1 - 2 \sin^2 \dfrac{\theta}{2} = 1 - \dfrac{2\theta^2}{4} = 1 - \dfrac{\theta^2}{2}$;

Therefore $l = 2\pi a + 2d\theta^2 + 2d\left(1 - \dfrac{\theta^2}{2}\right)$

$= 2\pi a + 2d + d\theta^2$

Call l' the value of l when $x = 0$, or $\theta = 0$,

$\therefore l' = 2\pi a + 2d$

$\therefore l - l' = d\theta^2$

But $2x = d \sin \theta = d\theta$

$\therefore \theta = \dfrac{2x}{d}$

$\therefore l - l' = d \cdot \dfrac{4x^2}{d^2} = \dfrac{4x^2}{d}$

which expresses the difference of the lengths in a convenient form.

It is apparent at once that l is greater than l'.

It has been stated that this difference is very trifling in many cases, and the following example is an illustration.

Let the diameters of two speed pulleys be 4, 6, 8, 10, 12 inches respectively, and let l be the length of strap for the extreme pair of 12 and 4, while l' is the length for the equal pair of 8 and 8, the distance between the centres of the pulleys being 6 feet.

Then $l - l' = 4 \dfrac{(6-4)^2}{72} = \dfrac{16}{72} = \dfrac{2}{9}$ inch, which is rather less than $\tfrac{1}{4}$ of an inch.

In practice, open bands are usually preferred to those which are crossed; the latter embrace a larger portion of the circumference, and are therefore less liable to slip, but they rub and wear away at the point where they cross.

160. After what has been stated it is scarcely necessary to point out the express use of *conical pulleys*; they form an obvious modification of step pulleys where the change is continuous instead of being abrupt.

There are two forms, one where the oblique edges of a section are parallel lines and the other where the convexity of one section exactly fits into the concavity of the other.

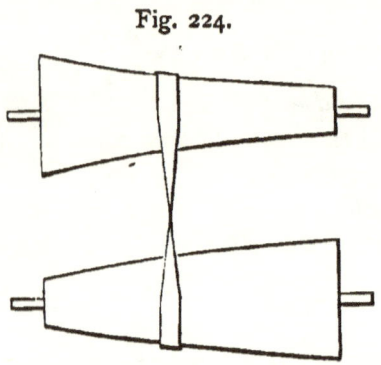

Fig. 224.

If the band be crossed we have seen that it will retain the same tension in every position upon the cones. If it be open, it will be less stretched at the middle than at either end according to Art. 159. When the obliquity is small the difference becomes absorbed in the elasticity or 'sag' of the band; otherwise it must be provided for by giving convexity to one or both of the cones.

The rotation of the upper cone being uniform, it is evident that the rotation of the lower cone will decrease as the strap is shifted towards the right hand.

One of the cones is sometimes replaced by a cylindrical drum, in which case the strap must be kept stretched by a tightening pulley.

As an illustration, we refer to the use of these conical pulleys in the manufacture of stoneware jars and other large earthenware vessels; where a mass of clay is fashioned into the required form upon a rotating table, and the workman varies the speed of the table according to the requirements of the work by shifting the driving strap along a pair of cones.

In cotton-spinning machinery these cones are of great service, and now that we are approaching the conclusion of our subject, it may be well to point out the manner in which certain simple elementary movements are combined together so as to form an entire machine, which before it is properly understood might appear to be very complex and intricate. For this purpose we will recur again to Houldsworth's differential motion, of which a drawing is given in Fig. 225, and which may be regarded as an exercise in the application of mechanism.

The operation of spinning, so far as it is carried on in the machine before us, is effected by passing a partially twisted fibre or *roving* through a tube, called a flier, attached to the end of a spindle, and then causing both the flier and the bobbin to rotate with a high velocity. Before the fibre reaches the fliers it is elongated or drawn out by a combination of rollers, moving at different speeds and called *drawing rollers*; it is therefore of necessity fed on at a fixed uniform rate.

The flier and the bobbin both rotate together and thus twist the roving, but they also rotate at somewhat different speeds, by which arrangement it is provided that the joint operations of twisting the thread and of winding it up upon the bobbin shall go on together.

A bobbin with its spindle and flier is shown in the sketch: it will be seen that the roving passes down through the hollow vertical arm and is carried to the bobbin by a finger; the finger is pressed against the bobbin by the centrifugal action of a small elongated piece which runs down the side of the arm, and which, by its tendency to get as far as possible from the axis of the spindle during its rotation, keeps the finger pressed against the surface of that portion of roving which is already wound upon the bobbin. This part of the apparatus has formed the subject-matter of a most lucrative invention.

As the winding goes on the spindle with its flier rises and falls, and winds the fibre in uniform layers.

Thus the spindle and flier rotate together, and they are driven by skew-bevils whereof one is shown at the bottom of the drawing.

The bobbin rotates independently of the spindle, and is also driven by skew-bevils, whereof one is shown just underneath the bobbin.

These bevil-wheels are in direct communication with the

spur-wheels marked 'to spindles' and 'to bobbins' in the drawing.

It will be understood that the winding on will take place when the spindles and the bobbins move at different velocities, and that either may go faster than the other. We shall take the case in which the bobbins precede the fliers.

Since the spindles with their fliers move at a fixed velocity, while the bobbins are continually filling with the rovings and becoming larger, we infer that the bobbins will require a smaller amount of rotation relatively to the fliers, in order that the winding up of the fibre, which is being fed on at a fixed rate by the drawing rollers, may take place uniformly. Hence, if the bobbin runs in advance of the flier the speed of revolution has to be diminished as its diameter becomes larger.

Refer now to the sketch, and it will be seen that the power may pass through the combination of bevil wheels to the three spur wheels placed in a line at the extremity of the 'driving axis' and connected with the cone marked as the 'driver'; the driving power then crosses over to the follower, and enters the combination of bevil wheels by the small pinion upon the axis of the lower cone which gears with the large spur wheel marked H, which latter wheel *rides loose* upon the driving axis.

The combination of four bevil wheels is exactly analogous to that discussed in Art. 116, the two wheels B and B are equivalent to a single wheel, and prevent the one-sided, unbalanced action which would otherwise occur.

The wheel A is fixed to the driving shaft, the wheel C rides loose upon it, but is fastened immovably to the spur wheel marked 'to bobbins,' the function of which has been already explained.

We have to prove that the combination of the two cones with the spur and bevil wheels is capable of gradually reducing the velocity of the bobbins as they fill up with the roving.

Differential Motion.

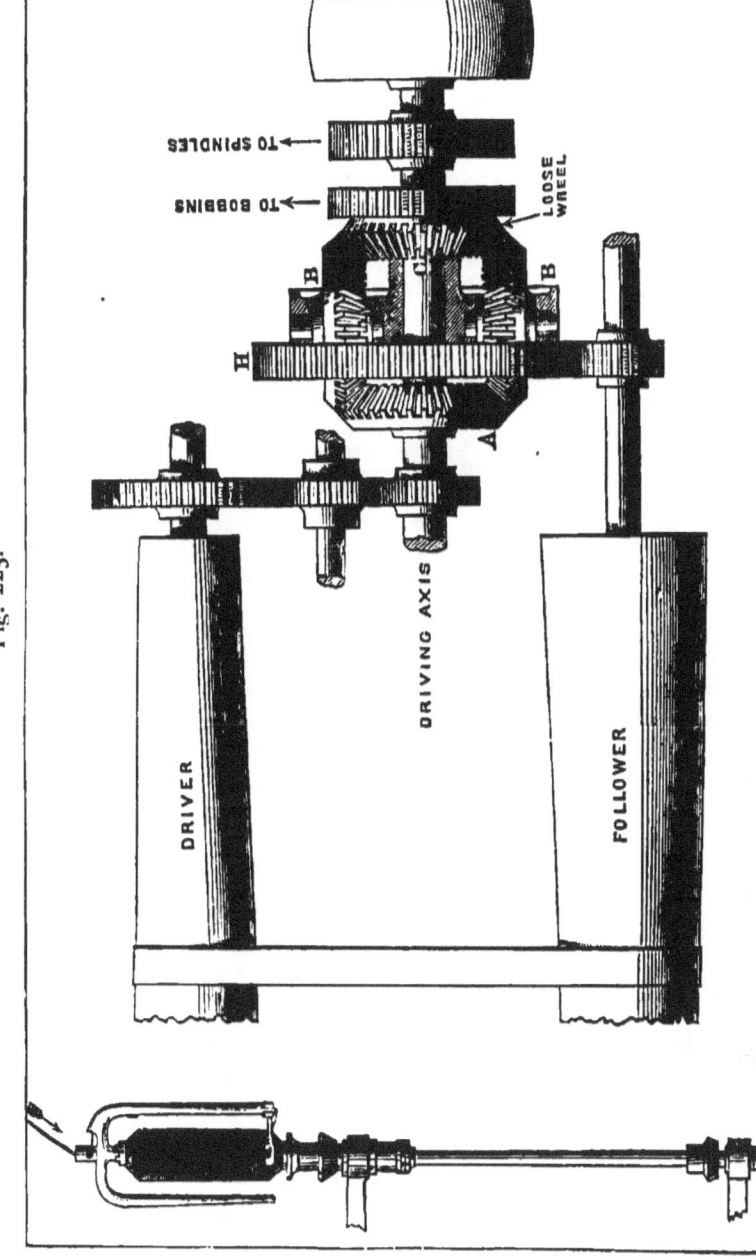

Fig. 225.

Assume that the cones are equal in section where the strap is placed, then the speed of the first cone will be reduced to $\frac{1}{2}$ by the combination of three spur wheels starting from the driving axis, and thus the pinion which drives H will move at $\frac{1}{2}$ the speed of the driving axis.

But H is five times as large as that pinion, hence the velocity of H is $\frac{1}{10}$th that of the driving axis.

The wheel H also rotates in the *opposite* direction to the driving axis.

Take now the formula $n = \dot{m}e + (1-e)a$. (See Art. 111.)

Here $e = -1$, since A, B, and C are equal,

and $a = -\dfrac{m}{10}$ as we have just shown;

$$\therefore n = -m + 2\left(-\dfrac{m}{10}\right)$$
$$= -m - \dfrac{m}{5}$$
$$= -\dfrac{6m}{5}.$$

Hence the speed of the bobbin pinion is to that of the flier pinion as 6 to 5, or 18 to 15. The negative sign merely showing that the loose wheel C revolves in the opposite direction to the driving axis.

The student may be surprised to find that all this apparently reducing arrangement has ended in making the last spur wheel in the train turn faster than the driving axis; but an explanation is found in the fact that the rotation of H takes place in the opposite direction to the driver, that is, in the same direction as the loose wheel C, and accordingly we shall find that if the velocity of H be reduced we shall also reduce the inequality between the velocity of the bobbins and spindles.

Conceive now that the strap is shifted towards the right hand until the sections of the cones are in the proportion of 2 to 3; that is, nearly as far as the spur wheels.

The velocity of H will be reduced two-thirds, and will become equal to $\frac{1}{15}$th that of the driving axis

$$\therefore n = -m + 2\left(-\frac{m}{15}\right)$$
$$= -m - \frac{2m}{15}$$
$$= -\frac{17m}{15},$$

or the relative speed of the bobbin pinion to that of the flier pinion is reduced from 18 to 15, and now stands at 17 to 15.

It is hoped that the complete action of the apparatus is now sufficiently explained, and there is only one refinement in construction which remains to be pointed out. It will be seen that the upper cone is slightly concave and the lower one convex : this configuration is adopted because the absolute increase in the diameter of a bobbin bears a ratio to the actual diameter which is not constant but is continually diminishing in a small degree. The mechanic must not forget or overlook any material point in working out his design.

161. The *star wheel* is used in cotton-spinning machinery, and is analogous to the Geneva stop.

Fig. 226.

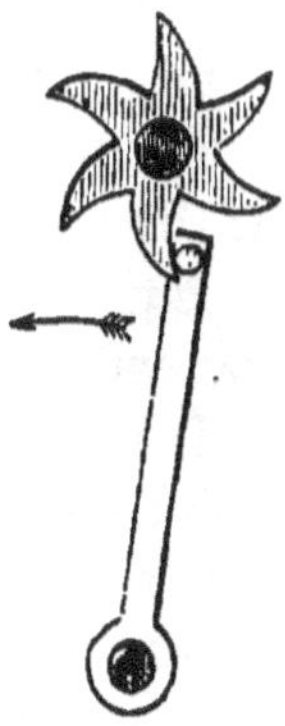

If the convex portion E were removed so as not to interfere with the rotation of A, we should virtually possess a star wheel in the disc B. (See Art. 155.)

In that case each rotation of A would advance B by the space of one tooth, or we should convert a continuous circular motion into one of an intermittent character.

The usual form of the star wheel is given in the sketch, where the revolving arm encounters and carries forward a tooth at each revolution. The action is the same as if a wheel with one tooth were to drive another with several teeth.

162. *Rolling curves* have been employed to vary the relative angular velocity of two revolving pieces.

The guiding proposition connected with this subject is the following:—

Prop. Where two curved plates, centred upon fixed axes, roll together, the point of contact must always lie in the line of centres.

Let two such plates be centred at A and C, and suppose that P and Q represent two points which will come together when the curves move each other by rolling contact.

Fig. 227.

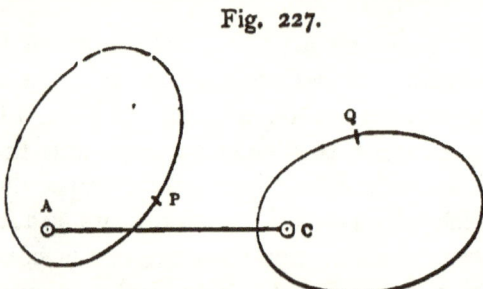

Then P describes a circle round A as the plate revolves, and Q describes a circle round C; hence P and Q will come into contact whenever these circles meet each other. Now P and Q only meet once in one revolution, and therefore the circles can only meet once, that is, they touch each other. But the circles can only touch in the line A C, therefore the point of rolling contact must lie in A C.

Cor. This is equivalent to saying that A P + C Q = a constant quantity.

Further than this, the curves will have a common tangent at the point where they roll upon each other.

Ex. Two equal ellipses which are centred on opposite foci will roll together.

It is the property of an ellipse that the tangent at any point P makes equal angles with the focal distances S P and H P, that is, that the angles S P T and H P *t* are equal at every point of the curve, and again, that the sum of the lines S P and H P is a constant quantity. (Fig. 228.)

The two equal ellipses centred upon opposite foci are represented in contact at P. (Fig. 229.)

Rolling Curves. 239

Let PT be the tangent to the ellipse A at P, and Pt the tangent to the ellipse B at P.

Fig. 228.

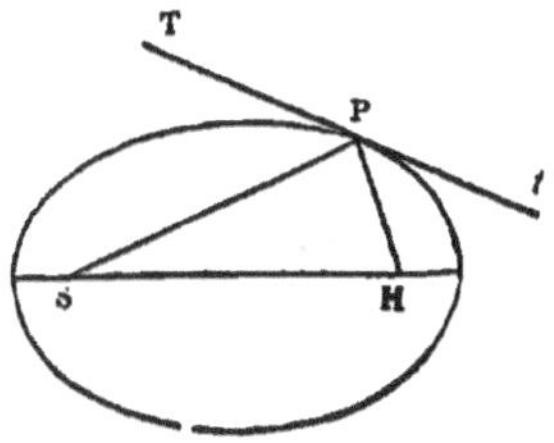

Fig. 229.

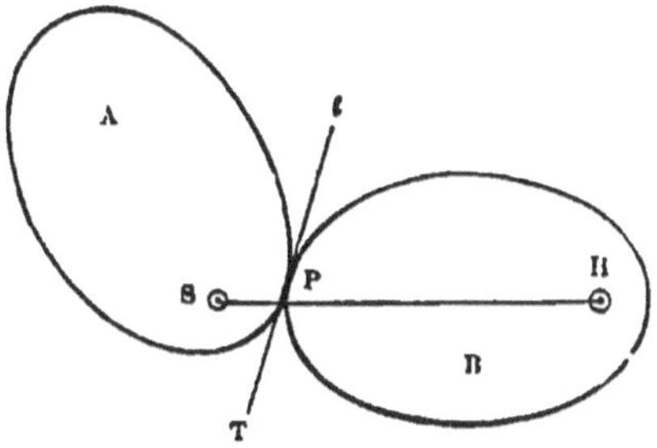

Then $SPT = tPH$ by the property above stated, ∴ TPt is a straight line, or the curves have a common tangent at the point P, also $SP + HP =$ a constant quantity, and the two conditions of rolling are fulfilled.

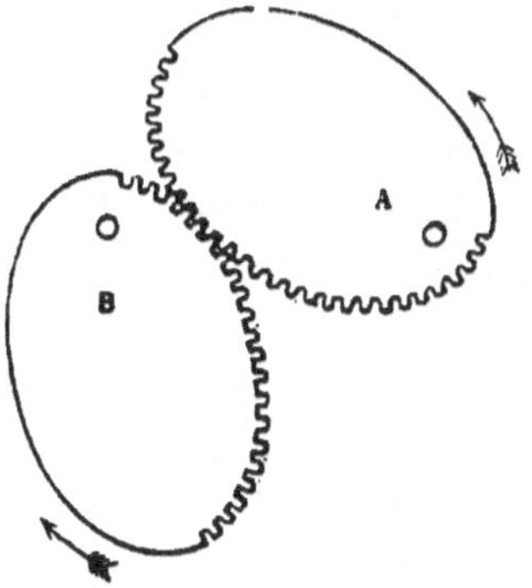

In practice rolling curves must be provided with teeth upon the retreating edge, otherwise the driver would leave the follower, and the revolution would not be completed. (Fig. 230.)

Fig. 230.

As is usual in all cases where segmental wheels are employed, a guide must direct the teeth to the exact point where they commence to engage each other.

The guide may be dispensed with by carrying the teeth all round the curves: this construction is usually adopted in practice, although, strictly speaking, it destroys the rolling action entirely.

163. A quick return of the table in small planing machines has been effected by the aid of rolling ellipses.

The table is driven by a crank and connecting rod, and the crank exists under the form of a flat circular plate, centred on one of the foci, and having a groove radiating from the axis as a line of attachment for one end of the connecting rod. As the plate may be set in any position upon the elliptical wheel, we propose to inquire what will be the effect of a change of direction in the groove or crank.

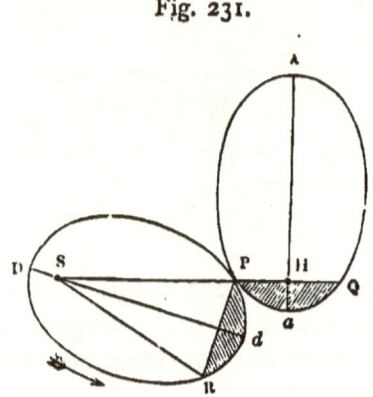

Fig. 231.

Let the ellipses have the position shown in Fig. 231, S and H being the centres of motion, and S P H Q being $\perp^r$ to A a, the axis of one ellipse. Draw P R $\perp^r$ D d, and let the ellipse D R d P be the driver, rotating in the direction of the arrow.

While P d R is rolling upon P a Q, the ellipse A a makes half a revolution; and while R D P is rolling upon Q A P, it makes the remaining half revolution.

Suppose D d to revolve uniformly, then the times of a half revolution of A a will be in the same proportion to each other as the angle P S R to the angle $360° -$ P S R. The quick half revolution occurs when the shaded segments are rolling upon each other. If, therefore, the table be made to move in the line H S produced, and the crank be placed in a direction $\perp^r$ A a, we shall obtain the greatest possible difference between the periods of advance and return.

The practical difficulty with rolling wheels exists during that part of the revolution where the driver tends to leave the follower, and it can only be obviated by making the teeth unusually deep; it is also essential that the wheels should work in a horizontal and not in a vertical plane.

164. An instance of rolling curves is exhibited in the sketch, and occurred in one of the many attempts made to improve the printing press before the invention of Mr. Cowper enabled the newspapers to commence a real and vigorous existence.

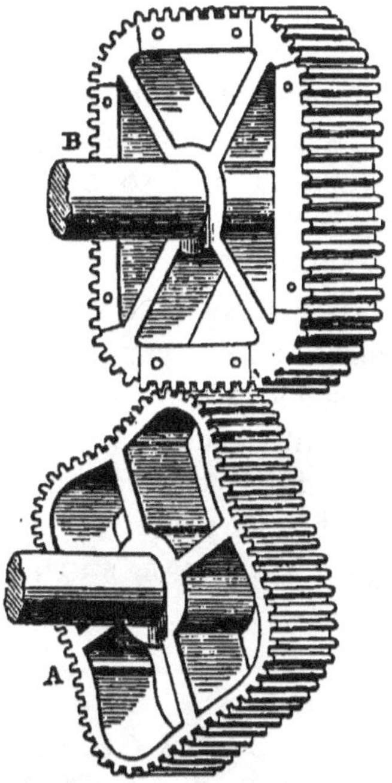

The type was placed upon each of the four flat sides of a rectangular prism to which the wheel B corresponded in shape, and the paper was passed on to a *platten* corresponding in form and size with the pitch-line of the wheel A.

The prism and platten being in the same relative position as the wheels B and A, we can understand that the type would be in the act of impressing the paper while the convex edge of the wheel A rolled upon the flat side of B, and that in this way we should obtain four impressions for each revolution of the wheels.

Fig. 232.

By this construction, the patentees, Messrs. Bacon and Donkin, intended to introduce the principle of continuous rotation as opposed to the reciprocating movement in a common press, and the object of imitating exactly upon the wheels A and B the form of the printing prism and of the platten, was to ensure that the paper and type should roll upon one another with exactly equal velocities at their opposing surfaces, and that no slipping or inequality of motion should destroy the sharpness of the impression.

165. The invention of *counting wheels* is due to the celebrated Cavendish, who constructed a piece of apparatus for registering the number of revolutions of his carriage wheel, which may be seen in the Museum of George III. at King's College.

There is but one guiding principle in this branch of mechanism, however varied may be the details of the separate parts.

Each wheel of a series, A, B, C, &c., possesses ten pins or teeth, and it is contrived that one tooth only of C shall be sufficiently long to reach those of B; similarly B is provided with one long tooth which is capable of driving A.

Thus C goes round ten times while B makes one revolution, and so on for the other wheels; in this way the series is adapted for counting units, tens, hundreds, thousands, &c.

In Fig. 233 the arm E F imparts rotation to the first, or *ratchet wheel*, by means of the paul H D; the number now registered is 988: after two vibrations of the arm the zero of C will reach the highest point, the tooth P will drive B through the space of one tooth: and the number registered will be 990; after ten more vibrations of the arm, P will again advance B, and at the same instant Q will move A,

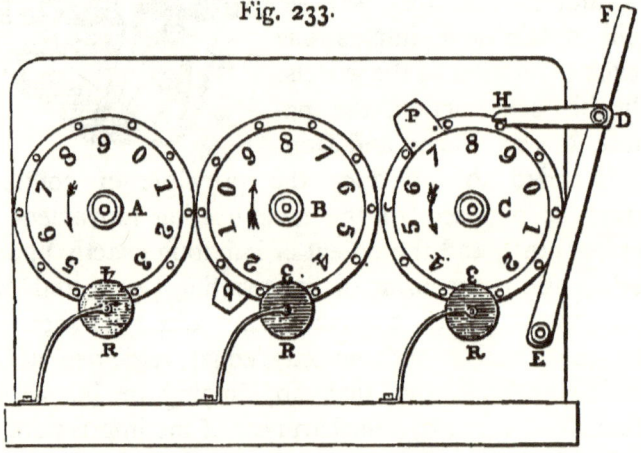

Fig. 233.

and will bring its zero up to the highest point: the three wheels will now register 000, having passed the number 999, which is the last they can give us.

A small counting apparatus is attached to every gas meter used in houses, and registers the number of cubic feet of gas consumed; here, however, the step by step motion is not employed, the dial plates are fixed, and a separate pointer travels round each dial respectively.

The pointers are placed upon the successive axes of a train of wheels, composed of a pinion and wheel upon each axis, the number of teeth on any wheel being ten times that upon the pinion which drives it. Suppose, for example, that the pointer on the plate registering thousands completes a revolution and adds ten thousand to the score, its neighbour on the left will have moved over one division on the dial registering tens of thousands, and thus an inspection of the pointers throughout the series will at once indicate the consumption of the gas.

These index-fingers move alternately in opposite directions, being attached to the successive axes of a train of wheels; the figures upon the counting wheels are also placed in the reverse order on every alternate wheel.

As we are only concerned with the counting apparatus, it is not necessary to explain the manner in which the flow of gas through the meter sets the train of wheels in motion, but we may point out that there is no ratchet wheel employed, and that the flow of gas keeps up a constant rotation in an endless screw, which starts the train and maintains it in action.

A reliable counting apparatus which will record the exact number of impressions made by a printing machine is indispensable in some public departments, and it is found that the best result is arrived at by combining a ratchet wheel having a few deep well-cut teeth with the train of wheels used in the gas meter.

The practice is to place upon the axis of the first or ratchet wheel carrying the units a pinion of 10 teeth gearing with a wheel of 100 teeth, then another pinion of 10 drives a wheel of 100 teeth, and so on as far as we please. The train of wheels cannot fail to record the hundreds, thousands, &c., accurately; the only possibility of a mistake occurs with the units, but if the paul be carried well over the teeth of the ratchet, and if the wheel itself be driven at each advance a little beyond the point necessary to give another unit, if, in other words, the movement should be a little over-pronounced, the register will be perfectly exact.

In order to avoid the objection that the successive wheels turn in opposite directions, an idle wheel is interposed between each alternate pinion of 10 and its wheel of 100. All the pointers then revolve in the same direction as the hands of a clock.

Where it is intended to print the figures registered, as in the numbering of bank notes, the step by step motion is essential, and further, each wheel must carry the letters upon its edge, and not upon the face; the apparatus employed is the same in principle as that of Cavendish, but the construction differs, the wheels being placed side by side and close together.

The construction of mechanism of this character has wonderfully advanced in late years, and many ingenious improvements have been originated; we do not intend to enter upon details of construction, but there is one contrivance known as a *masked wheel* which ought to be understood.

The object of the masked wheel is to enable a single machine to print the same number twice before the unit advances, as in numbering a cheque and its counterfoil.

The masked ratchet wheel consists of two ratchet wheels placed side by side, the teeth in one being the ordinary saw-cut teeth, while the teeth in the other are alternately shallow and deep; the main feature is that the bottom of the cut in a shallow tooth is farther from the centre than the apex of a tooth in the second wheel, whereby it follows that when the paul rests in the shallow cut it is quite clear of the teeth in the other wheel.

The two wheels ride upon the same axis, but move independently of each other, the ordinary ratchet wheel being connected with the numbering apparatus.

In the drawing a pin wheel represents the first ratchet wheel, and the second wheel C has its teeth in pairs, every alternate tooth being cut deeper in the manner stated.

Counting Wheels.

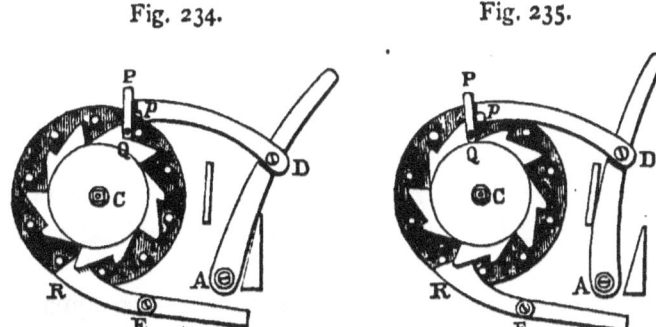

Fig. 234.   Fig. 235.

The paul P D has a pin Q, which must always engage with a tooth of the wheel C, but which may or may not drop low enough to take up a pin on the other wheel.

In Fig. 234 the paul has engaged a shallow tooth in the wheel C, and is ready to drive that wheel onwards; but the paul has its point p clear of the pins on the shaded wheel, being masked or prevented from acting by the shallowness of the cut at Q.

After the paul has driven on the wheel C by the space of a tooth, it will next fall into the deeper cut, when both Q and p will engage their respective teeth, and the two wheels will advance together in the manner shown in Fig. 235.

Thus two strokes of the paul are required in order to advance the pin wheel by one tooth.

Again, it is often an advantage to print the odd numbers, as 101, 103, 105, &c., in one column, and the even numbers, as 102, 104, 106, &c., in another column.

Or the same machine may be required to print consecutive numbers.

The arrangement for effecting this double purpose is a very simple one. The numbering wheels are carried by an arm in a circular sweep from the inking apparatus to the printing table; in travelling along they encounter the paul, which is fixed to the framework, and if the circle should simply graze, as it were, against the paul during its travel, one tooth only would be taken up; whereas by setting the paul so that it meets the circle at a point nearer its centre, and strikes it more directly, two teeth may be taken up, and thus either *one* or *two* units may be advanced at each impression.

166. *The differential worm wheel and tangent screw* is an ingenious combination which will now be understood without any drawing.

We remark that one principal feature in the apparatus which we have been examining consists in the reduction of motion caused by the single driving tooth. Now, it has been already stated that an endless screw is equivalent to a wheel with one tooth; we shall therefore be prepared to find that an endless screw is useful in mechanism of this character, and we may apply it as the driver instead of using a spur wheel.

Here two worm wheels, differing by one tooth in the number which they carry, are placed side by side and close together, so as to be capable of engaging with an endless screw. As the wheels are so very nearly alike, the endless screw can drive them both at the same time, and it is evident that one wheel will turn relatively to the other through the space of the extra tooth in a complete revolution, and that a very slow relative motion will thus be set up.

In this way, if one wheel carries a dial plate, and the other a hand, we may obtain the record of a very large number of revolutions of the tangent screw.

167. The problem of communicating a *uniform rotation* between two parallel axes, without the aid of wheelwork or the equivalent pulleys and bands, presents some useful exercises in mechanism.

The first idea of the student would probably be to place two equal cranks, one upon each axis, and to connect them by a rod whose length was equal to the distance between the axes.

Parallel Axes.

Let A P, B Q represent two such cranks, connected by a link P Q which is equal to A B, then A P Q B forms a parallelogram so long as P Q remains parallel to A B, and therefore A P and B Q will always be in similar positions so long as P Q remains parallel to itself. The parallelism of P Q is therefore the condition which ensures the joint rotation of the two cranks.

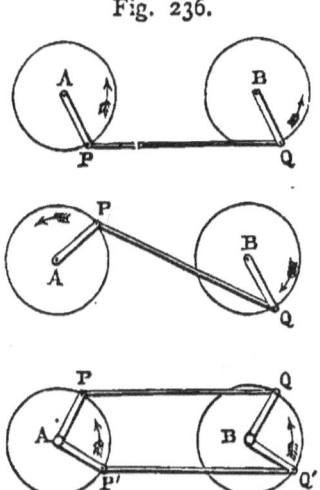

Fig. 236.

When the driving crank P Q comes upon the line of centres, the joint A P Q will bend if there be any resistance to the motion in the follower, and the arm B Q will then return while A P alone continues its rotation.

In that way one crank would continue to go round in a circle and the other would oscillate through 180°; this state of things is shown in the second diagram.

But the difficulty is easily overcome by employing a second combination exactly like the first, and making P A P' and Q B Q' bell-crank levers set at any convenient angle.

In this way, when P Q is passing through the dead points, P' Q' will hold it in a parallel position, and each connecting rod will prevent the other from taking that oblique position which is destructive of the required motion.

This is the principle of the coupling link between the two driving wheels in a locomotive engine. There are always two links, one on either side of the engine, and the cranks are of course at right angles.

The necessity of the second pair of cranks, with their link, is obvious upon a little consideration, and may be made very clear by constructing a small model of the arrangement; it is only necessary to make the links move in different planes so that they may be able to pass each other.

We next observe that any point in P Q will describe a circle equal to either A or B so long as P Q remains parallel to itself, and hence that a third crank equal to either A P or B Q, and placed between them, would be driven by P Q, and would further prevent P Q from getting into an oblique position at the dead points, or would produce the same result as the second pair of cranks with the link in the locomotive engine.

Again, the same would be true of any point R in a bar S R connected rigidly in any way with P Q, the point R would describe a circle equal either to A or B so long as the link P Q remained parallel to A B.

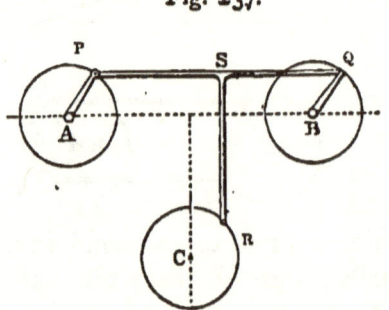

Fig. 237.

Also, if a crank were supplied at C R, the three cranks would go round together and P Q would remain parallel to itself.

168. Conceive now that three equal cranks, P p, Q q, R r, are centred at equal distances along a circle P Q R, as shown in Fig. 238, and let a second circle $p\,q\,r$, equal to P Q R, be jointed to the cranks at the points p, q, r.

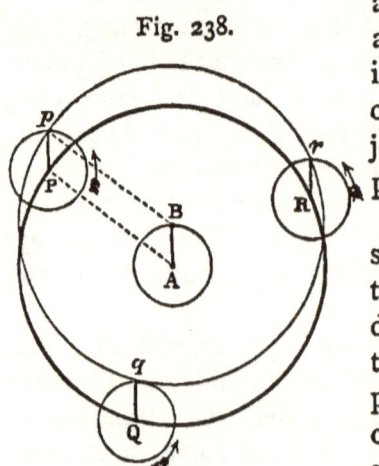

Fig. 238.

If the circle $p\,q\,r$ be shifted so that the cranks are allowed to rotate, each of them will describe a circle, the respective cranks will always remain parallel to each other, and the circle $p\,q\,r$ will move in such a manner that any line drawn upon it remains always parallel to itself.

Parallel Axes.

Hence the circle $p\,q\,r$ may be employed as a driver to rotate all three cranks at the same time, and while doing so, it will itself sweep round without the slightest movement of rotation upon its own centre.

It has what is sometimes called a motion of circumduction, and moves like the wheel C in Art. 110.

169. A very small alteration in the construction will give a combination which has been useful in rope-making machinery, and which was the first movement suggested for feathering the floats of paddle-wheel steamers.

Let the centres of the two circles P Q R and $p\,q\,r$ be made fixed centres of motion, and let P p, Q q, R r, remain as before.

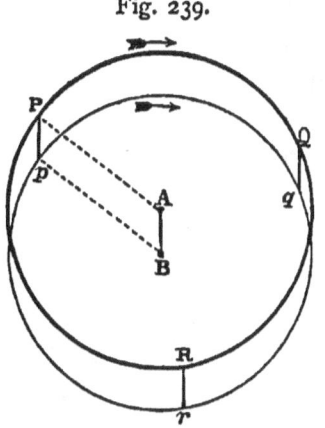

Fig. 239.

A power of rotation will now be given to both the circles, and P p, Q q, R r, will be the connecting links which always remain parallel to A B, the line of centres. That is to say, the rotation of the circle P Q R, about the centre A, will cause an equal rotation in the circle $p\,q\,r$, about its centre B, and P p, Q q, R r, must remain parallel to A B. It is the same combination that we started with, under a different aspect, by reason that the proportionate size of the pieces has been changed. The two circles A and B have been enlarged and brought together so that their circumferences overlap, and A P, B p are the parallel cranks.

So far as P p, Q q, R r, are concerned, they give us three supports for the bobbins or spools in rope making and the arrangement is a direct equivalent for the wheelwork in the ʻCordelier' described in Art. 116.

Regarding the contrivance as a method of feathering the floats in paddle wheels, we find that, in the year 1813,

Mr. Buchanan patented a form of paddle wheel in which one circle, as P Q R, carried the floats, and another circle, *p q r*, rotating with the former held these floats always in a vertical position, and so made them enter the water edgeways instead of striking it obliquely with the flat surface, as is the case in an ordinary paddle wheel.

170. This wheel of Buchanan has not been used for very sufficient reasons. It is not a good arrangement for the floats to enter the water in an exactly vertical line, because the motion of the vessel must compound with that of the floats, and the supposed vertical path will not be one in reality, any more than it would be in the case of a stone dropped from the same vessel into the water. The stone appears to fall in a vertical line, but is really projected forwards.

According to this view the float should enter the water

Fig. 240.

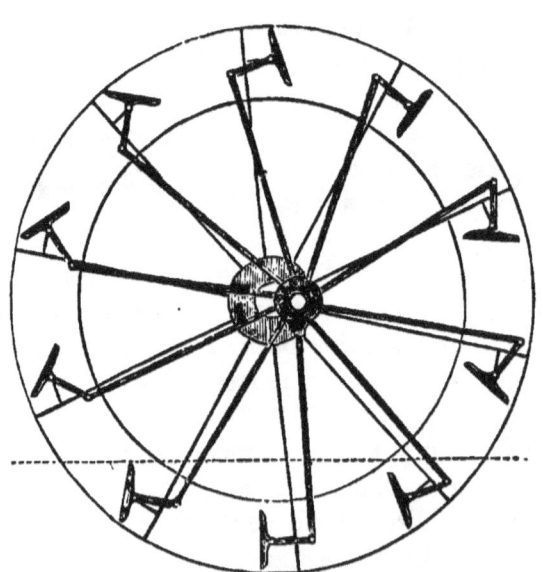

at an angle such that its line of direction will pass through the highest point of the wheel, this being the direction of the resultant of the two *equal velocities* impressed upon a point in the wheel, the one being that due to the vessel, the other being due to the rotation of the wheel.

This result may follow very closely from the construction in the drawing, which represents Morgan's wheel, where the floats are connected by rods with a ring that rotates round a fixed centre in the paddle-box; the floats are attached to small cranks and pivoted upon centres, one of them (the lowest in the drawing) being driven by a rigid bar which springs from a solid ring. Each float passes the lowest point in a vertical position, and is somewhat inclined when entering or leaving the water.

171. *Hooke's Joint* is a method of connecting two axes, whose *directions meet in a point*, in such a manner that the rotation of one axis shall be communicated to the other.

Fig. 241.

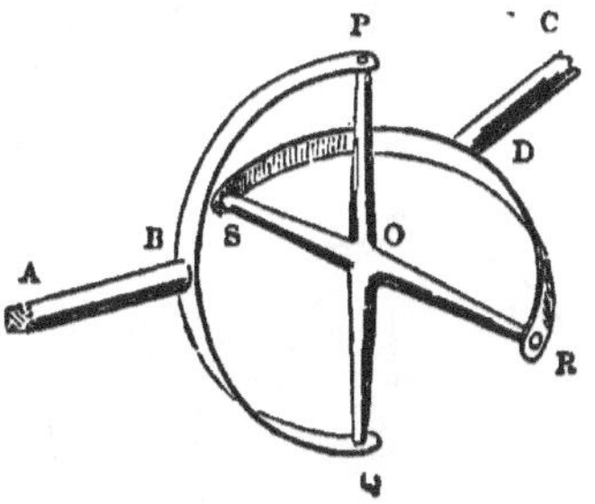

Here A B and C D represent two axes whose directions meet in the point O; the extremities of A B and C D terminate in two semicircular arms which carry a cross, P Q S R; the arms of this cross are perfectly equal, and the joints P, Q, S, and R permit the necessary freedom of motion.

As the axis A B revolves, the points P and Q describe a circle whose plane is perpendicular to A B, and at the same time the points S and R describe another circle whose plane is perpendicular to C D.

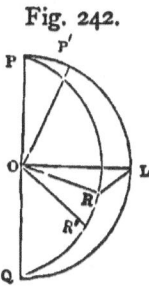

Fig. 242.

These two circles are inclined at the same angle as the axes, and are represented in Fig. 242; thus the arm O P starts from P and moves in the circle P P' L, while the arm O R starts from R and describes the circle R R' Q inclined to the former.

Let O P′, O R′ be corresponding positions of the two arms, then P′ R′ is constant, but changes its inclination at every instant, and as a consequence the relative angular velocities of O P′ and O R′ are continually changing.

To find the relative angular velocities of the axes A B and C D, we proceed as follows:—

Let the circle *p r q* (Fig. 243) represent the path of P, *p t q* being the projection upon this circle of the path of R, and suppose α to be the angle between A B and C D; then the dimensions of the curve *p t q*, which will be an ellipse, can be at once deduced from the equation $O\,t = O\,r \cos \alpha$.

Draw R *m* $\perp^r$ O *r*, then R *m* will be the actual vertical space through which O R has descended while O P describes the ∠ *p* O P. But the path of R is really a circle, and only appears to be an ellipse by reason of its being projected upon

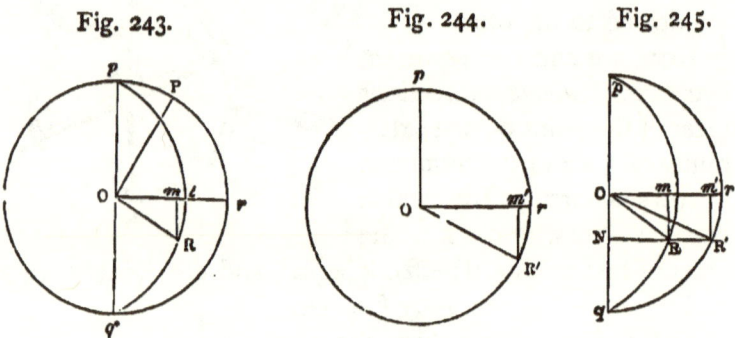

Fig. 243.　　　　Fig. 244.　　　　Fig. 245.

a plane inclined to its own plane; in order, therefore, to estimate the actual angular space through which O R has moved, we must refer this motion to the circle which R really describes (Fig. 244), and thus by making R′ *m*′ = R *m*, we can infer that *m*′ O R′ will be the angle which O R describes while O P moves through the angle *p* O P.

But the ∠ *p* O P = ∠ *m* O R, and hence we can represent the motion of both axes upon one diagram by combining the ellipse *p* R *q* with the circle *p r* R′ *q* (Fig. 245). This being done, we may draw R′ R N $\perp^r$ *p* O *q*, and join O R,

O R′; it will at once appear that the angles R O r, R′ O r are those described in the same time by the axes A B and C D.

Hence the axes A B and C D revolve together, but unequally, and the angles which they describe in the same time can always be found by construction.

First draw the circle p r q in a plane perpendicular to one axis and having o for its centre, next construct the ellipse whose major axis is the diameter p o q equal to P O Q, and whose minor axis is the product of P O Q × the cosine of the angle between the axes. Then take O R′ any position of O P, draw R′ R N perpendicular to p o q, join O R′ and O R, it now appears that R′ O r and R O r will represent the angles described by the axes A B and C D in the given time.

Furthermore, O R and O R′ coincide when R is at the end of an axis of the ellipse p R q, an event which must happen four times as the cross goes round once; and there is therefore this curious result, that however unequal may be the rate at which the axes are at any time revolving, they will coincide in relative position four times in one revolution.

The single joint may often be very useful in light machinery which is required to be moveable, and the parts of which do not admit of very accurate adjustment; but it will be understood that the friction, and especially that irregularity which we have just proved to exist, would render it necessary to confine the angle between the shafts within narrow limits in actual practice.

172. Now that the general character of the movement is understood we shall be in a position to comprehend the change which is effected by interposing a *double joint* between the axes.

1. Take the case where A B and C D are parallel axes connected by an intermediate piece B C, having a Hooke's joint at both the points B and C.

Conceive that the arms of the crosses at B and C are placed in the manner shown in the sketch, or let each vertical arm be connected with the forks at B and C.

If A B revolves uniformly, B C will also revolve with a varying velocity dependent upon the angle A B C, but the variable velocity which B C receives from A B is precisely the same as that which it would receive from D C if the latter axis were the driver and were to revolve uniformly.

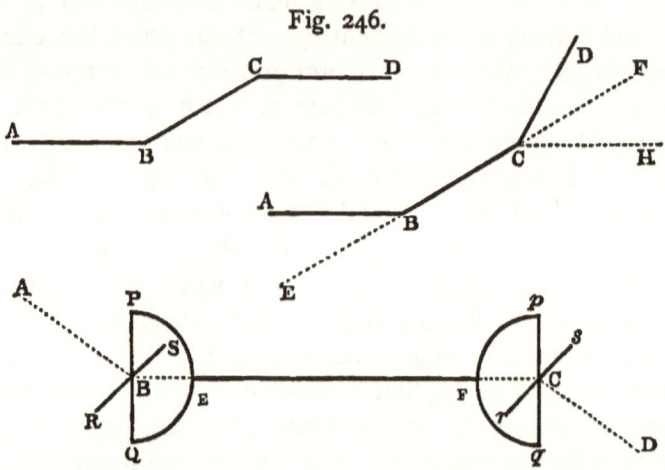

Fig. 246.

It follows therefore that the motion which A B imparts to C D will be a uniform velocity of rotation exactly equal to that of A B.

Hence a double Hooke's joint may be used to communicate uniform motion between two parallel axes whose directions nearly coincide.

If, however, the construction were varied and the vertical arm P Q of the first cross were connected with E, while the horizontal arms *s*, *r* were connected with F, we should communicate no doubt a motion of rotation between the axes, but it would no longer be uniform but variable, by reason that we could not return by the same course reversed under like conditions. The deviations from uniform rotation would no longer oppose and correct each other, but they would act together and increase the inequality. This is seen at once upon constructing the diagrams which represent the relative rotation between each pair of axes.

2. Let A B and C D be inclined to each other, and be connected by the piece B C jointed at B and C, and so placed that the angle A B C is equal to the angle B C D.

As in the former case we must be careful to connect B and C with the corresponding arms of the crosses, and we have seen that the inequality produced by D C in the motion of C B depends both upon the angle B C D and the position of the cross; it is therefore the same whether C D lies in the direction shown or in the dotted line C H parallel to A B. In both cases the angle between the axes and the position of the cross will respectively coincide.

But we have seen that when the parallel axes A B and C H are connected by joints at B and C in the manner stated, the axes A B and C H will rotate with equal uniform velocities, and we conclude, therefore, that they will also rotate in a similar manner when placed in the position A B C D.

Hence a double Hooke's joint may be employed to communicate a uniform rotation between two axes inclined at a given angle.

173. We pass on to an entirely different method of connecting *two parallel axes*.

Suppose A and B to represent two axes, and D E, F H to be a rigid rectangular cross whose arms always pass through the points A and B.

It is a geometrical fact that if D E revolves uniformly about A, then F H will also revolve uniformly about B.

The motion will only be possible so long as the arms of the cross have perfect liberty to slide through the points A and B as well as to rotate about them.

Let this be arranged, and join A B; then we have

P A B + P B A = 90° in every position of the cross.

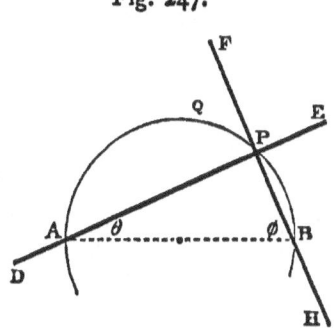

Fig. 247.

Hence if the angle P A B increase by the rotation of P A, the angle P B A must diminish equally by the rotation of P B, or E D and F H must revolve with equal angular velocities.

Also P, the angle of the cross will describe a circle whose diameter is A B, and our proposition follows directly from Euclid, for if P move on to any point Q, the angles Q A P, Q B P are angles in the same segment of a circle and are therefore equal to each other.

This movement was put into a practical shape by Mr. Oldham, and used in machinery at the Bank of England. The student may easily construct a model after the manner of Hooke's Joint, when the centre of the cross will be seen to describe a circle whose diameter is the perpendicular distance between the axes while the arms of the cross slide to and fro through holes in the forked arms that spring from the axes and support them.

174. Another method of connecting two axes by the aid of a cross possesses the property of causing one axis to rotate twice as fast as the other.

Let an arm P C Q be centred at its middle point C and drive the rectangular cross which is centred at R by means of pins at P and Q running in grooves along F H and D E.

Describe a circle with centre C and radius equal to C P or C R; draw any fixed diameter A C B, and join A R.

Then it is proved in Euclid that the angle at the centre of a circle is double the angle at the circumference when both angles stand upon the same arc,

that is ∠ P C A = 2 ∠ P R A or the angular velocity of C P is twice that of the cross.

As the driver P C Q revolves the pins P and Q will oscillate to and fro along their respective grooves and will traverse through the centre R.

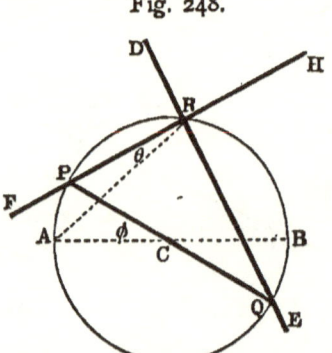

Fig. 248.

Bell-crank Levers.

175. *Bell-crank levers* serve to change the line of direction of some small motion, and are of universal application. They consist simply of two arms standing out from a fixed axis so as to form a bent lever.

1. Suppose it to be required to construct a bell-crank lever so as to change the direction of some small motion from the line B D into the line D A, where B D and D A meet in a point D.

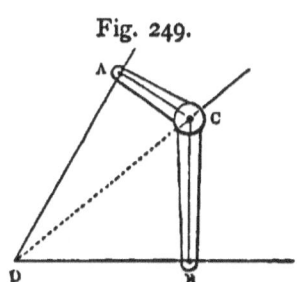

Fig. 249.

Draw D C, dividing the angle at D into two parts whose *sines are in the ratio of the velocities* of the movements in the given directions.

This may be done by setting up perpendiculars anywhere on B D and D A in the required ratio, and drawing straight lines through their extremities parallel to B D and D A respectively. These parallel lines will intersect somewhere in D C, and will determine that line.

In D C take any convenient point c, and draw C A, C B, $\perp^r$ to D A and D B respectively, then A C B will be the bell-crank lever required.

This is the construction, and it can be immediately verified, for the arcs described by the extremities A and B, when the lever A C B is shifted through a small angle a, will be represented by A C × a and C B × a respectively, and will measure the velocities of the points A and B.

Hence $\dfrac{\text{velocity of A}}{\text{velocity of B}} = \dfrac{A C \times a}{C B \times a} = \dfrac{A C}{C B} = \dfrac{\sin C D A}{\sin C D B}$.

It is evident that the movements in D A and D B are very nearly rectilinear, and will become more so the further we remove c from the point D.

Any play which may be necessary at the joints A and B, by reason that the ends of the levers really describe small circular arcs, may be easily provided for in the actual arrangement.

2. To change the direction from one line to another not intersecting it.

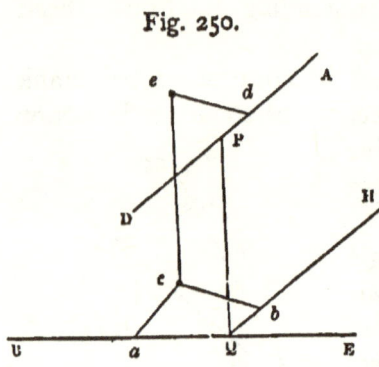

Fig. 250.

Draw P Q, a common perpendicular to the lines A D and B E (Fig. 250); through Q draw Q H parallel to D A; construct a bell-crank lever, *a c b*, for the movements as transferred to the lines B Q, Q H; draw *c e* parallel to P Q and equal to it, and further make *e d* parallel and equal to *c b*.

The piece *a c e d* will be the lever required; what has been done is this, a bell-crank lever *a c b* has been formed by the rule given above in order to transfer the motion from B E to Q H, and then the motion in Q H has been shifted into another line D A parallel to itself.

An example of the use of a bell-crank lever of the second kind occurs in that portion of the rifling machine discussed in Art. 37, and by means of it the longitudinal movement of the rod R N is transferred to another horizontal line P Q, lying at some little distance above R N.

176. We purpose now, in conclusion, to examine a little more particularly the mechanism of an escapement, so as to gain some idea of the refinements of its construction when applied to the best made clocks or watches, and we will review very rapidly the elementary facts which are to be found in the books on mechanics.

An imaginary or *simple pendulum* is a conception of mathematicians, and is defined as being a single particle of matter, P, suspended by a string, D P, without weight.

This particle may swing to and fro in a *vertical plane* under the action of the pull of the earth, and the *oscillation* of the pendulum is the whole movement which it makes in one direction before it begins to return, viz. A C B.

The Pendulum.

The *time of a small oscillation* is the period of this movement, and is given by the formula :—

$$t = \pi \sqrt{\frac{l}{g}},$$

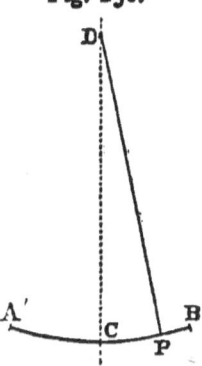

Fig. 251.

where t = time of a swing in seconds,
l = length of the string in feet,
g = 32·2 feet.

Make $t = 1$ and we have the length of a pendulum oscillating seconds, which is a little less than a mètre, being equal to 39·14 inches.

The discovery of the so-called pendulum law was made by Galileo, who noticed that a lamp swinging by a chain in the metropolitan church at Pisa made each movement in the same time, although gradually oscillating through smaller arcs while coming to rest.

For ordinary purposes the time of a swing may be supposed to depend only on the length of the pendulum and not at all upon the arc through which it swings. It is practically very nearly the same, for arcs up to 2 or 3 degrees on each side of the vertical, and it may be shown by calculation that the error introduced by assuming the law to be strictly true, in the case of a pendulum moving through an arc of 5 degrees from the vertical position would only amount to $\frac{1}{2000}$th part of the time of a swing.

A seconds' pendulum in a well-made clock swings through about 2 degrees on each side of the vertical.

The question now arises, how is the law of the swing of this imaginary pendulum to be applied in the regulation of clocks, and wherein does a solid heavy pendulum, which we must of necessity employ, resemble or differ from an ideal pendulum?

Conceive a straight uniform bar of iron to be suspended close to one end upon small triangular wedges of hardened steel, technically called knife edges, which rest upon perfectly horizontal agate or steel planes, and then to be set swinging; it is evident that each particle of the bar will endeavour to observe the pendulum law stated above, and will tend to swing in different times. But all the particles must swing together, and the result is that a sort of compromise takes place between the different tendencies, and the whole bar swings as if its material were all collected into one dense point at a certain distance from the knife edge, which is known as the *centre of oscillation*. Thus the solid pendulum swings as an ideal pendulum would do whose length was exactly equal to the distance between the centres of suspension and oscillation.

This is again the theory of the *rigid pendulum*, as used by clockmakers, which has been investigated by Hughens, and by others after him, and has led to many interesting experiments in applied mechanics.

We should here point out that in the case of the ideal pendulum, where there is no artificial resistance and no friction, the movement is in theory perpetual. In the case of our rigid pendulum the friction at the point of suspension and the resistance of the air would gradually diminish the arc of swing, and the movement would slowly subside and die away, although it might be many hours before it became quite imperceptible. The mechanism of a clock must therefore so act upon the pendulum as to maintain its swing unimpaired by these resistances. In the article on escapements we stated generally the function both of an escape wheel and a pendulum, and it should be understood that in applying the pendulum as part of the apparatus of an anchor escapement, it is usual to connect it with the anchor by means of a fork so that both swing together as one piece, although each has a separate point of suspension.

Any impulse or check given to the anchor is therefore an impulse or check felt at once upon the pendulum.

Refer now to the recoil escapement described in Art. 16, and conceive that the escape wheel is urged onwards in the

Recoil Escapement.

direction of the arrow by the force of the clock train, so as to press its teeth slightly against the pallets of the anchor, the pendulum being hung from its point of suspension by a thin strip of steel, and vibrating with the anchor in the manner already stated.

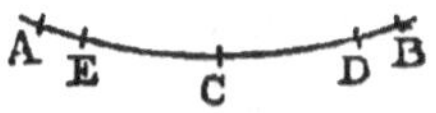

Let the arc A E C D B be taken to represent the arc of swing of the bob of the pendulum.

Fig. 252.

As the pendulum moves from B to E the point q of the escape wheel rests upon the oblique surface A m of the pallet, and presses the pendulum onward until the point of the tooth escapes at the end of the pallet. For an instant the escape wheel is free, and tries to fly round, but a tooth is caught at once upon the opposite side by the oblique edge B n, and the escape wheel then presses against the pendulum and tends to stop it, until finally the pendulum comes to rest at the point A, and commences the return swing.

What now has been the action? From B to E the force of the train as existing in the escape wheel has been acting with the pendulum and has performed its proper office in assisting to maintain the swing; whereas from E to A this force has acted *against* the pendulum.

So also on the return swing, the escape wheel will act *with* the pendulum from A to D, and *against* it from D to B.

The action then is alternately *with* and *against* the pendulum, and it might be supposed that the injurious effect of a pressure against the pendulum would be entirely corrected by the maintaining force in the other part of the swing; but this is not the case, the pendulum no longer moves with what we may call its natural swing, as a free pendulum would oscillate, and any variation in the maintaining force will disturb the rate of the clock.

The matter has been carefully analysed by mathematicians, and they have shown that the principle of this escapement is radically bad, because it is impossible to remedy entirely the harm which is done by continually interfering with the swing of the pendulum.

There occurs also the useless expenditure of work. It is almost superfluous to remark that no mechanical arrangement will ever bear scrutiny when it is so constructed as to throw away work.

177. The *dead-beat escapement* was invented by Graham and at once removes this primary objection.

It is given in the sketch, and the student will observe that the pallet A has its lower edge in the form of a circular arc A q whose centre is C, and again that the upper portion of the pallet B is also a circular arc struck about the same centre. The oblique surfaces $q\,m$, $n\,p$ complete the pallets. Take the case shown in the diagram, which is enlarged so as to make the action more apparent. As long as the tooth is resting on the circular portion $n\,r$ of the pallet the pendulum is free to move and the escape wheel is locked. Hence in

Fig. 253.

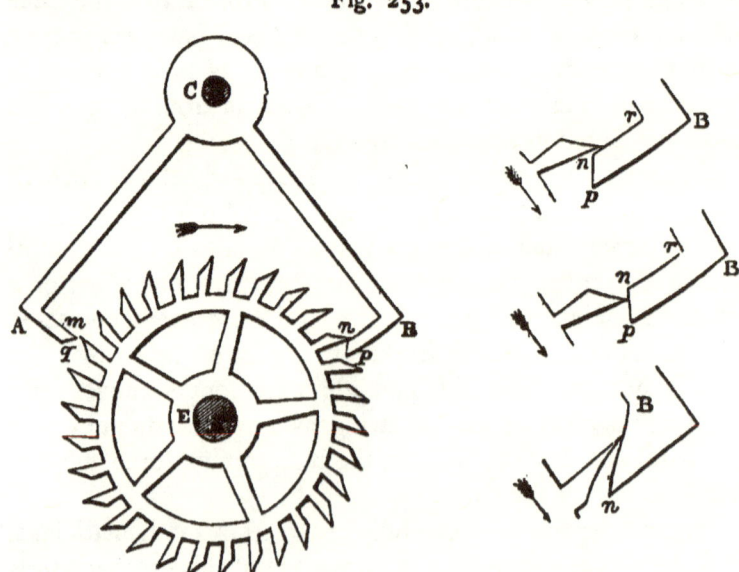

the portion E A and back again through A E, there is no

Dead-beat Escapement. 263

action against the pendulum except the very minute friction which takes place between the tooth of the escape wheel and the surface of the pallet; through a space E C D the point of the escape wheel is pressing against the oblique edge np and is urging the pendulum forward.

Then at D the tooth upon the opposite side falls upon A q and the escape wheel is locked; from D to B, and back again to D, there is the same friction which acted through E A or A E; whereas from D to E, the point of a tooth presses upon qm and urges the pendulum onward, at E another tooth is locked upon the pallet B n, and thus the action is reproduced in the order in which it has been described.

It follows that any action against the pendulum is eliminated or, more correctly, is rendered as nearly as possible harmless, and the difference between the 'recoil' and the 'dead beat' will be understood upon contrasting the three enlarged diagrams which sufficiently explain themselves, the lower sketch referring to the recoil escapement.

The term 'dead beat' has been applied because the seconds' hand which is fitted to the escape wheel stops so completely when the tooth falls upon the circular portion nr. There is none of that recoil or subsequent trembling which occurs when a tooth falls upon B n and is driven back.

The actual construction of the dead-beat escapement having been explained, it only remains for us to state two of the principal conclusions which follow from a theoretical inquiry into the motion of the pendulum.

1. All action against the pendulum should be avoided, and if some such action be inevitable, it should at any rate be reduced to the smallest amount that is practicable.

2. The maintaining force should act as directly as possible, and the impulse should be given through an arc which is bisected by the middle point of the swing.

That is, the arc of impulse D C E should be bisected at the lowest point C.

This latter condition cannot be exactly fulfilled, because

the point of the escape wheel must fall a little beyond the inclined slope of the pallet in order that it may be locked with certainty.

178. We have next to show that these principal conditions obtain also in the construction of the escapements of watches, and that the principle of 'dead beat' is recognised in the three forms which are in common use.

And here we may remark that the pendulum of a clock appears as the *balance wheel* in a watch.

A wheel, pivoted on very small steel pivots, and working in jewelled supports, is attached to a flat spiral steel spring in pocket watches, or to a more powerful helical steel spring in marine chronometers. This wheel vibrates under the action of the pull of the spring just as a pendulum would do under the pull of the earth, but under better conditions theoretically, for the force of the spring increases with the angle through which the balance wheel swings, and in direct proportion to that angle; the result therefore is that the swing of the watch pendulum is always performed in very nearly equal times whether the arc of swing be increased or diminished.

We have what is technically called an *isochronous pendulum* in the balance wheel, and this is important, because the time is not affected by small changes in the arc of swing. Further, it should be noted that the balance wheel swings through an angle which is enormous as contrasted with the swing of the pendulum, being more than a whole revolution in the case of the chronometer or lever watch.

Consider now the construction of the chronometer escapement which fulfils our conditions with an exactness that may well surprise us, and which exhibits in its arrangement a marvellous amount of mechanical skill and forethought.

The detent, which corresponds to the anchor pallets, consists of four principal parts :—

1. The locking stone, D, a piece of ruby, upon which the tooth of the escape wheel rests.

Chronometer Escapement. 265

2. The discharging spring, A r, which is a very fine strip of hammered gold.

3. A screw at A to fix A r to the stem of the piece s D.

4. The shank of the detent, consisting of a projecting arm D q, the part D A, and the portion at s, which is cut away to form a spring which may bend and act as a pivot on which the whole detent can be moved a little.

The small circle on the left hand has a projecting piece which keeps the escapement in action, and it is a part of the stem of the balance wheel.

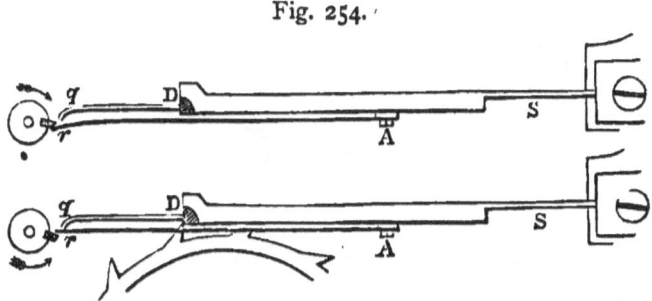

Fig. 254.

As the balance wheel swings to and fro, this roller also vibrates, and when passing downwards it encounters the spring A r, and pushes it aside without any perceptible effort, because the spring bends from the distant point A.

On its return the projection finds the spring to be capable of bending, not from the distant point A, but only from the point q against which it rests; the roller therefore takes the spring and the whole detent with it and raises the locking stone D from the point of the escape wheel, the escape wheel at once flies round, and before it can be caught upon the next tooth by the return of the detent to its normal position is enabled to give an impulse to the balance wheel by striking against the point d in the manner shown in Fig. 255.

The whole arrangement can now be studied from the drawings, and is complete with the exception of a banking

266   *Elements of Mechanism.*

screw which supports the detent when coming back to the position of rest; it will be seen that the large circle F is fixed to the smaller one, and that the projection marked *d* is quite clear of the escape wheel while a tooth is resting against the detent.

Fig. 255.

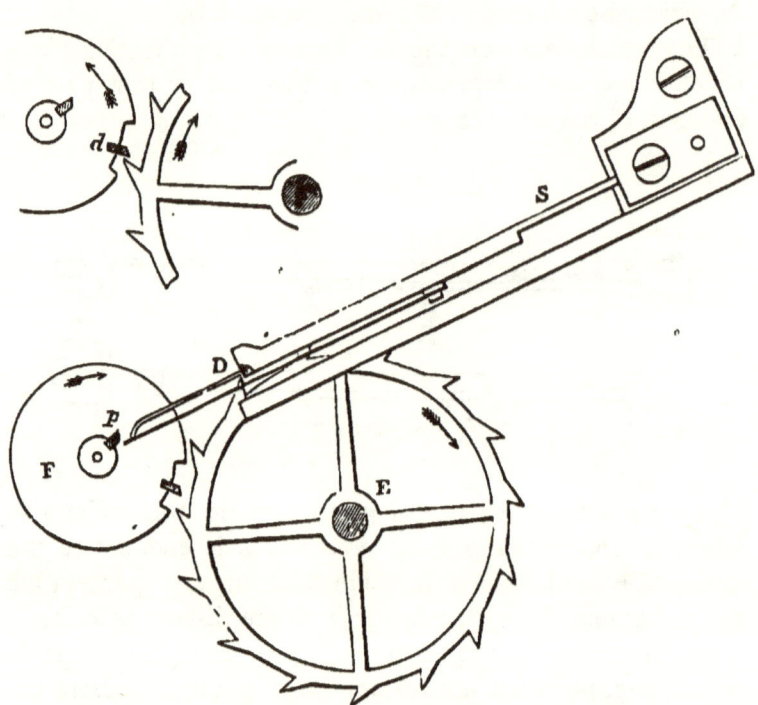

The advance of the escape wheel is so instantaneous that it is not seen to move, it appears to tremble a very little, but it comes to rest again so quickly that the eye cannot follow and can scarcely detect the motion. It is of course made evident by watching the spokes of the wheel.

What then has been the action? In the first swing of the balance the only obstacle has been the bending aside of the spring A*r*, which is no more than bending a light feather. In the second swing the pendulum or balance wheel has had to lift the detent; this is a momentary and very small action against it, but as quick as thought the action is compensated and the balance receives its impulse through equal distances on each side of the middle of its swing according to the principle of the dead-beat escapement.

Here, then, theory and practice are in exact accordance.

It should be noted that the impulse is given at every alternate swing of the balance, and not with every swing as in the case of the clock pendulum.

179. The *Lever Escapement* comes next in order, and here we return to the anchor pallets. The escape wheel is locked by these pallets and gives its impulse upon their oblique edges in the manner described in Art. 177.

The balance wheel is free during the greater part of the swing (hence the name of *Detached Lever*), and oscillates through an angle of quite 400°. The unlocking occupies an angle of about 3°, and the impulse is given through about 9°.

These are just the conditions which prepare us for the principle of the dead beat.

Fig. 256.

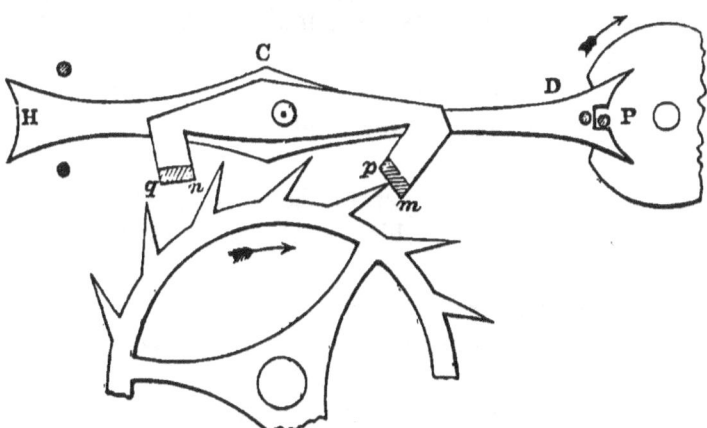

The pallets $m\,p$, $q\,n$ are jewels inlaid into the arms, the light steel bar D H is the lever moveable about C as a centre; an open jaw at one end is capable of receiving a ruby pin attached to the roller F which is on the axis of the balance wheel and moves with it. There are also banking pins and a small guard pin to prevent the lever from falling out of position.

As the balance vibrates the pin P swings to and fro with it; in doing so it enters the open cut at the end of the lever and removes the locking portion of the pallet from the point of the escape wheel. Instantly the escape wheel flies forward and by pressing against one oblique edge suddenly pushes on P, and the lever is no longer moved by the balance wheel but imparts an impulse to it. Very soon, that is after the 9° of the swing are consumed, the ruby leaves the lever behind and the wheel goes on detached and unchecked in its swing.

On its return the pin finds the lever where it had left it, carries it forward, unlocks the escape wheel, receives its impulse, leaves the lever behind, and the balance is free for the rest of the swing.

The only action against the balance is that of unlocking the escape wheel so as to enable it to give the impulse; this is very brief in duration as compared with the whole swing, and the watchmaker takes care that it shall be as little as possible. The impulse is given just at the middle of the vibration, and the construction follows out the theory very closely.

180. Lastly, we may refer to the escapement of the so-called *Geneva Watches*, which is Graham's cylinder movement.

Here the balance is attached to a very thin cylinder centred at o, and the point of a tooth rests upon either the outside or the inside of this cylinder during a part of the swing; in this respect the action corresponds exactly to the friction of the escape tooth against the circular part of the pallets in the dead beat.

As the cylinder vibrates round its centre o, the tooth $p\,n$ comes under the edge at r, and pushes the cylinder onward, this gives an impulse; the tooth soon passes *r*, flies into the cylinder and is stopped by the concave surface near *s*; the cylinder now vibrates in the opposite direction, *p n* escapes, and in doing so gives another impulse at *s* to the cylinder in the opposite direction, and thus the action goes on.

Fig. 257.

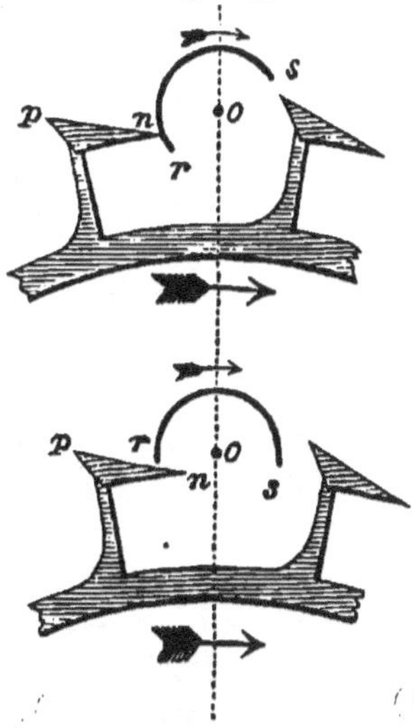

The impulse would not be given in the middle of the swing but through small arcs equally distant from the middle point and equal in length to each other. Hence this combination is nearly identical with the dead-beat escapement, although inferior to it in this latter particular.

The manner in which the effects of the expansion and contraction of the material of the pendulum rod and the balance wheel, due to changes of temperature, are rendered innocuous, forms a separate branch of the subject.

LONDON: PRINTED BY
SPOTTISWOODE AND CO., NEW-STREET SQUARE
AND PARLIAMENT STREET

TEXT-BOOKS OF SCIENCE, MECHANICAL AND PHYSICAL,
ADAPTED FOR THE USE OF ARTISANS AND OF STUDENTS IN PUBLIC AND SCIENCE SCHOOLS.

Now in course of publication, in small 8vo. each volume containing about Three Hundred pages,

A SERIES OF
ELEMENTARY WORKS ON MECHANICAL AND PHYSICAL SCIENCE,
FORMING A SERIES OF
TEXT-BOOKS OF SCIENCE,
ADAPTED FOR THE USE OF ARTISANS AND OF STUDENTS IN PUBLIC AND OTHER SCHOOLS.

The first Thirteen of the Series, edited by T. M. GOODEVE, M.A. Barrister-at-Law, Lecturer on Applied Mechanics at the Royal School of Mines; and the remainder by C. W. MERRIFIELD, F.R.S. an Examiner in the Department of Public Education, and late Principal of the Royal School of Naval Architecture and Marine Engineering, South Kensington.

THE Reports of the Public Schools Commission and of the Schools Inquiry Commission, as well as the evidence taken before several Parliamentary Committees, have shewn that there is still a want of a good Series of TEXT-BOOKS in Science, thoroughly exact and complete, to serve as a basis for the sound instruction of Artisans, and at the same time sufficiently popular to suit the capacities of beginners.

Messrs. LONGMANS & Co. have accordingly made arrangements for the issue of a Series of Elementary Works in the various branches of Mechanical and Physical Science suited for general use in Schools, Colleges, and Science Classes, and for the self-instruction of Working Men.

These books are intended to serve for the use of practical men, as well as for exact instruction in the subjects of which they treat; and it is hoped that, while retaining that logical clearness and simple sequence of thought which are essential to the making of a good scientific treatise, the style and subject-matter will be found to be within the comprehension of working men, and suited to their wants. The books will not be mere manuals for immediate application, nor University text-books, in which mental training is the foremost object; but are meant to be *practical treatises, sound and exact in their logic, and with every theory and every process reduced to the stage of direct and useful application, and illustrated by well-selected examples from familiar processes and facts.*

TEXT-BOOKS, NOW PUBLISHED, EDITED BY T. M. GOODEVE, M.A.

THE ELEMENTS OF MECHANISM.
Designed for Students of Applied Mechanics. By T. M. GOODEVE, M.A. Barrister-at-Law, Lecturer on Mechanics at the Royal School of Mines. New Edition, revised; with 257 Figures on Wood. Price 3s. 6d.

METALS, THEIR PROPERTIES AND TREATMENT.
By CHARLES LOUDON BLOXAM, Professor of Chemistry in King's College, London; Professor of Chemistry in the Department of Artillery Studies, and in the Royal Military Academy, Woolwich. With 105 Figures on Wood. Price 3s. 6d.

INTRODUCTION TO THE STUDY OF INORGANIC CHEMISTRY.
By WILLIAM ALLEN MILLER, M.D. LL.D. F.R.S. late Professor of Chemistry in King's College, London; Author of 'Elements of Chemistry, Theoretical and Practical.' New Edition, revised; with 71 Figures on Wood. Price 3s. 6d.

ALGEBRA AND TRIGONOMETRY.
By the Rev. WILLIAM NATHANIEL GRIFFIN, B.D. sometime Fellow of St. John's College, Cambridge. Price 3s. 6d.

NOTES ON THE ELEMENTS OF ALGEBRA AND TRIGONOMETRY;
With SOLUTIONS of the more Difficult QUESTIONS. By the Rev. WILLIAM NATHANIEL GRIFFIN, B.D. sometime Fellow of St. John's College, Cambridge. Price 3s. 6d.

PLANE AND SOLID GEOMETRY.
By the Rev. H. W. WATSON, formerly Fellow of Trinity College, Cambridge, and late Assistant-Master of Harrow School. Price 3s. 6d.

THEORY OF HEAT.
By J. CLERK MAXWELL, M.A. LL.D. Edin. F.R.SS. L. & E. Professor of Experimental Physics in the University of Cambridge. New Edition, revised; with 41 Woodcuts and Diagrams. Price 3s. 6d.

TECHNICAL ARITHMETIC AND MENSURATION.
By CHARLES W. MERRIFIELD, F.R.S. an Examiner in the Department of Public Education, and late Principal of the Royal School of Naval Architecture and Marine Engineering, South Kensington. Price 3s. 6d.

KEY TO MERRIFIELD'S TEXT-BOOK OF TECHNICAL ARITHMETIC AND MENSURATION.
By the Rev. JOHN HUNTER, M.A. one of the National Society's Examiners of Middle-Class Schools; formerly Vice-Principal of the National Society's Training College, Battersea. Price 3s. 6d.

ON THE STRENGTH OF MATERIALS AND STRUCTURES.
The Strength of Materials as depending on their quality and as ascertained by Testing Apparatus. The Strength of Structures as depending on their form and arrangement, and on the Materials of which they are composed. By JOHN ANDERSON, C.E. LL.D. F.R.S.E. Superintendent of Machinery to the War Department. Price 3s. 6d.

ELECTRICITY AND MAGNETISM.
By FLEEMING JENKIN, F.R.SS. L. & E. Professor of Engineering in the University of Edinburgh. New Edition, revised. Price 3s. 6d.

WORKSHOP APPLIANCES.
Including Descriptions of the Gauging and Measuring Instruments, the Hand-Cutting Tools, Lathes, Drilling, Planing, and other Machine Tools used by Engineers. By C. P. B. SHELLEY, Civil Engineer, Hon. Fellow and Professor of Manufacturing Art and Machinery at King's College, London. With 209 Figures on Wood. Price 3s. 6d.

PRINCIPLES OF MECHANICS.
By T. M. GOODEVE, M.A. Barrister-at-Law, Lecturer on Applied Mechanics at the Royal School of Mines. With 208 Figures and Diagrams on Wood. Price 3s. 6d.

TEXT-BOOKS, NOW PUBLISHED, EDITED BY C. W. MERRIFIELD, F.R.S.

QUANTITATIVE CHEMICAL ANALYSIS.
By T. E. THORPE, F.R.S.E. Ph.D. Professor of Chemistry in the Andersonian University, Glasgow. With 88 Figures on Wood. Price 4s. 6d.

Text-Books of Science.

INTRODUCTION TO THE STUDY OF ORGANIC CHEMISTRY;
The CHEMISTRY of CARBON and its COMPOUNDS.
By HENRY E. ARMSTRONG, Ph.D. F.C.S. Professor of Chemistry in the London Institution. With 8 Figures on Wood. Price 3s. 6d.

QUALITATIVE CHEMICAL ANALYSIS AND LABORATORY PRACTICE.
By T. E. THORPE, Ph.D. F.R.S.E. Professor of Chemistry in the Andersonian University, Glasgow; and M. M. PATTISON MUIR, F.R.S.E. With Plate and 57 Figures on Wood. Price 3s. 6d.

TEXT BOOKS PREPARING FOR PUBLICATION, TO BE EDITED BY T. M. GOODEVE, M.A.

ECONOMICAL APPLICATIONS OF HEAT.
Including Combustion, Evaporation, Furnaces, Flues, and Boilers.
By C. P. B. SHELLEY, Civil Engineer, and Professor of Manufacturing Art and Machinery at King's College, London. With a Chapter on the Probable Future Development of the Science of Heat, by C. WILLIAM SIEMENS, F.R.S.

THE STEAM ENGINE.
By T. M. GOODEVE, M.A. Barrister-at-Law, Lecturer on Mechanics at the Royal School of Mines.

SOUND AND LIGHT.
By G. G. STOKES, M.A. D.C.L. Fellow of Pembroke College, Cambridge; Lucasian Professor of Mathematics in the University of Cambridge; and Secretary to the Royal Society.

TEXT-BOOKS PREPARING FOR PUBLICATION, TO BE EDITED BY C. W. MERRIFIELD, F.R.S.

TELEGRAPHY.
By W. H. PREECE, C.E. Divisional Engineer, Post Office Telegraphs; and J. SIVEWRIGHT, M.A. Superintendent (Engineering Department) Post Office Telegraphs.

RAILWAY APPLIANCES.
Including Permanent Way, Points and Crossings, Stations and Station Arrangements, Signals, Carriage and Waggon Stock, Breaks, and other Details of Railways.
By J. W. BARRY, Member of the Institution of Civil Engineers, &c. With Woodcuts.

PRACTICAL AND DESCRIPTIVE GEOMETRY, AND PRINCIPLES OF MECHANICAL DRAWING.
By C. W. MERRIFIELD, F.R.S. an Examiner in the Department of Public Education, and late Principal of the Royal School of Naval Architecture and Marine Engineering, South Kensington.

ELEMENTS OF MACHINE DESIGN.
With Rules and Tables for Designing and Drawing the Details of Machinery. Adapted to the use of Mechanical Draughtsmen and Teachers of Machine Drawing.
By W. CAWTHORNE UNWIN, B.Sc. Assoc. Inst. C.E. Professor of Hydraulic and Mechanical Engineering at Cooper's Hill College.

PHYSICAL GEOGRAPHY.
By the Rev. GEORGE BUTLER, M.A. Principal of Liverpool College; Editor of 'The Public Schools Atlas of Modern Geography.'

** *To be followed by other works on other branches of Science.*

London, LONGMANS & CO.

www.ingramcontent.com/pod-product-compliance
Lightning Source LLC
Chambersburg PA
CBHW031936230426
43672CB00010B/1944